TAMING TIME, TIMING DEATH

Taming Time, Timing Death

Social Technologies and Ritual

Edited by

DORTHE REFSLUND CHRISTENSEN
Aarhus University, Denmark

RANE WILLERSLEV
University of Oslo, Norway

Routledge
Taylor & Francis Group
LONDON AND NEW YORK

First published 2013 by Ashgate Publishing

Published 2016 by Routledge
2 Park Square, Milton Park, Abingdon, Oxfordshire OX14 4RN
711 Third Avenue, New York, NY 10017, USA

First issued in paperback 2016

Routledge is an imprint of the Taylor & Francis Group, an informa business

British Library Cataloguing in Publication Data
Taming time, timing death : social technologies and ritual.
1. Death—Social aspects. 2. Time—Sociological aspects.
3. Time perspective. 4. Funeral rites and ceremonies.
I. Christensen, Dorthe Refslund. II. Willerslev, Rane, 1971–
306.9–dc23

The Library of Congress has cataloged the printed edition as follows:
Christensen, Dorthe Refslund.
Taming time, timing death : social technologies and ritual / by Dorthe Refslund Christensen and Rane Willerslev.
p. cm.
Includes bibliographical references and index.
ISBN 978-1-4094-5068-9 (hardback)
1. Death—Social aspects. 2. Time perspective. 3. Time—Sociological aspects.
4. Funeral rites and ceremonies. I. Willerslev, Rane, 1971– II. Title.

HQ1073.C477 2013
306.9–dc23

2012040432

ISBN 13: 978-1-138-25352-0 (pbk)
ISBN 13: 978-1-4094-5068-9 (hbk)

Contents

List of Figures

Notes on Contributors

Mikkel Bille holds a PhD in anthropology from University College London (2009). His doctoral work was on material culture among the Bedouin in Jordan. He is currently working on sustainable lighting technologies relating to cultural notions of atmosphere in both Jordan and Denmark. He has edited *An Anthropology of Absence* (2010) with Tim Flohr Sørensen and Frida Hastrup.

Kristian Bjørkdahl is Research Fellow at the University of Oslo, Norway, where he is writing a dissertation on neopragmatism and human–animal relations. He is a researcher at the Centre for Development and the Environment (SUM) at the same institution. His background is in interdisciplinary cultural studies, and his main research interest is Richard Rorty's pragmatism, book history and reception studies, and the rhetoric of the animal and environmental movements.

Dorthe Refslund Christensen, MA, PhD, is Associate Professor in the Department of Aesthetics and Communication at Aarhus University, Denmark. Her research interests include the ritualization and mythologization of transformation in performative identity strategies and everyday life practices, as well as off- and online community building.

Michael Hviid Jacobsen is Professor of Sociology and Director of the Master's Programme in Humanistic Palliative Care at Aalborg University, Denmark. He has written and edited many books within the sociology of death and dying, most recently *Deconstructing Death* (2012). Besides this he works on sociological theory and is inspired particularly by the work of Zygmunt Bauman and Erving Goffman. He is also a founder and member of the board of directors of the Nordic Network of Thanatology.

Michael C. Kearl is Professor of Sociology at Trinity University, USA, where he has taught for 35 years. Author of *Endings: A Sociology of Death & Dying* (Oxford University Press, 1989), he has published extensively on such subjects as the political uses of the dead in civil religion, the centrality of abortion and euthanasia in America's culture wars, the increasing roles of the dead in popular culture, growing old in death-denying cultures, and American immortalism and its battles against extinction.

Murray Last is Professor Emeritus in the Department of Anthropology, University College London. In 1967 he published *The Sokoto Caliphate*, which has now been published also in Hausa as *Daular Sakkwato*. He is co-editor (with G.L. Chavunduka) of *The Professionalisation of African Medicine* (1986) and has published over a hundred articles on northern Nigeria.

Stuart McLean is Associate Professor of Anthropology at the University of Minnesota. He has carried out fieldwork in Ireland and, more recently, the Orkney Islands. His research interests include historical and philosophical anthropology, the relationship between the living and the dead and exploring possibilities of creative collaboration between anthropology, art and literature. His publications include *The Event and its Terrors: Ireland, Famine, Modernity* (2004).

Lotte Meinert is Professor of Anthropology in the Department of Culture and Society at Aarhus University, Denmark. She leads a research project on recovery after war in Northern Uganda, as well as a Center for Cultural Epidemics (EPICENTER). She has carried out long-term fieldwork in Uganda since 1993 and is the author of *Hopes in Friction: Schooling, Health and Everyday Life in Uganda* (2009).

Henrik Hvenegaard Mikkelsen is a PhD student in anthropology at Aarhus University and Moesgaard Museum, Denmark, where he also completed his MA. His interests include personhood, violence and egalitarianism, which will be explored in his dissertation based on recent, extended fieldworks among the Bugkalot of the Philippines. His research concerns the relation between cosmology and masculinity and how cosmology is inscribed in the male body, along with the social and surrounding territorial world.

Rune Nyord was trained as an Egyptologist at the University of Copenhagen and since 2010 he has held the Lady Wallis Budge Fellowship at Christ's College, Cambridge. His research focuses on ancient Egyptian language and religion. He is the author of *Breathing Flesh: Conceptions of the Body in the Ancient Egyptian Coffin Texts* (2009) and co-editor (with Annette Kjølby) of *'Being in Ancient Egypt': Thoughts on Agency, Materiality and Cognition* (2009).

Jesper Østergaard, PhD, is an associate professor in the Department of Culture and Society at Aarhus University, Denmark. Trained in the Study of Religion and Cognitive Semiotics, Østergaard's research includes empirical studies of Tibetan religion and pilgrimage and theoretical developments in the intersection between religion, cognition and landscape.

Knut Rio is Professor of Social Anthropology at the University of Bergen, Norway, and is responsible for the ethnographic collections at the University Museum of Bergen. He has worked on Melanesian ethnography since 1995, with fieldwork in Vanuatu. His work on social ontology, production, ceremonial exchange, witchcraft and art in Vanuatu has resulted in journal publications and the monograph *The Power of Perspective: Social Ontology and Agency on*

Ambrym Island, Vanuatu (2007). He has also co-edited *Hierarchy: Persistence and Transformation in Social Formations* (with Olaf Smedal, 2009), *Made in Oceania: Social Movements, Cultural Heritage and the State in the Pacific* (with Edvard Hviding, 2011) and *The Arts of Government: Crime, Christianity and Policing in Melanesia* (with Andrew Lattas, 2011).

Kjetil Sandvik, MA, PhD, is an Associate Professor in the Department of Media, Cognition and Communication at the University of Copenhagen, Denmark. His research interests include digital aesthetics, storytelling as a format in strategic communication, new media and experience economies, cross-media communication and computer games.

Karen Lykke Syse is an agronomist and ethnologist and holds a PhD in cultural history. Her research interests pivot around the histories and ideologies of nature, focusing on environmental discourse and practice; agrarian and arboreal landscape studies; social and cultural aspects of hunting and angling and nature in early modern thought. Her ongoing projects are on contemporary relationships to animals as food.

Piers Vitebsky is Head of Anthropology and Russian Northern Studies, Scott Polar Research Institute, University of Cambridge and Professor at the University of Tromsø, Norway. He studied ancient languages before becoming an anthropologist researching religion, psychology and ecology in Arctic Siberia and Tribal India. His books include *The Reindeer People: Living with Animals and Spirits in Siberia* (2005); *Shamanism* (1995); and *Dialogues with the Dead: The Discussion of Mortality among the Sora of Eastern India* (1993), while his documentary films include *Siberia – After the Shaman* (1991).

Susan Reynolds Whyte is a Professor in the Department of Anthropology at the University of Copenhagen, Denmark. She is the author of *Questioning Misfortune: The Pragmatics of Uncertainty in Eastern Uganda* (1997) and co-author of *Social Lives of Medicines* (2003). She has co-edited *The Social Context of Medicines in Developing Countries* (1988); *Disability and Culture* (1995); *Disability in Local and Global Worlds* (2007) and *Generations in Africa: Connections and Conflicts* (2008).

Rane Willerslev is Professor of Anthropology and Director of the Museum of Cultural History at the University of Oslo, Norway. He has his PhD from the University of Cambridge. His main field of research has been hunting and spiritual knowledge among Siberia's indigenous peoples. He is the author of *Soul Hunters: Hunting, Animism and Personhood among the Siberian Yukaghirs* (University of California Press, 2007) and *On the Run in Siberia* (University of Minnesota Press, 2012). He is currently the leader of a three-year interdisciplinary research project funded by the Danish Research Councils, exploring the relationship between death, materiality and time.

Preface

Eventually we all die – but more importantly, we experience death head-on, when someone close to us dies. The present book series *Studies in Death, Materiality and the Origin of Time* stipulates that this fact constitutes the origin of human conceptions of time. Time permeates everything, but except for time itself all things are perishable. Yet, it is only through the perishable world of things and bodies that we sense time. The book series explores the fact that human experiences and conceptions of time inherently hinge on the material world, and that time as a socially experienced phenomenon cannot be understood as separate from material form or expression. As such, the series departs from a persistent current within Western thinking. Philosophy, biology and physics among other disciplines have studied time as an essential, ethereal and abstract concept. In the same way, death has often been conceived of in abstract and sometimes transcendental terms as occupying one extreme margin of human life. As an alternative, this series suggests that bodily death and material decay is a central point of reference in social life, which offers key insights into human perceptions of time. The series brings together scholars from anthropology, ethnology, archaeology, sociology, media study, cultural study, Egyptology and religion, all of whom explore the topic on the basis of their particular disciplinary focus and in a range of different empirical present and historical settings.

This first volume explores the meaning which death has for human understandings of time across cultures, stretching from the ancient Egypt and Nepal over tribal India, the Arabic world and the Philippines to North Siberia, East and West Africa, and various parts of the Western world. The aim of this grand interdisciplinary and cross-cultural venture is to explore the various 'social technologies' used, including timework, ritual, objects and interpersonal relations, when confronted with death.

Series editors
Rane Willerslev, Museum of Cultural History, University of Oslo
Dorthe Refslund Christensen, Aarhus University

Acknowledgements

This book would not have been possible without the help, support and input of several institutions and colleagues. We gratefully acknowledge the support from the Independent Research Councils of Denmark's Sapere Aude programme for young research leaders, which made it possible for us to establish an interdisciplinary research group, which for three years could explore the interface between death and time in different cultural settings. We are extremely grateful to the small circle of friends and colleagues in both Aarhus and Oslo, who by participating regularly in our so-called 'Magic Circle' meetings, have helped developing the overall argument of the book along with many of its chapters. This book is dedicated to the memory of Robert Hertz, who made us understand, more than anyone, how time can be thought in relation to death.

Introduction

Rane Willerslev, Dorthe Refslund Christensen and Lotte Meinert

This book suggests that bodily death and material decay are a central point of reference that offers key insights into human perceptions of time. We explore the way human beings handle death and decay by means of social technologies, including timework, ritual and material objects, which help establish human temporalities.

The event of death demands, in a concrete material and social sense, that other human beings deal with the body of the deceased. Not at any time, but *in time*, which is a deadline dictated by the corporeality of the dead, decomposing body. In his classic essay on death from 1907, philosopher and sociologist of religion Robert Hertz made it clear that the imperative to deal with the dead body is of a social kind:

> We see life vanish but we express this fact by the use of a specific language: it is the soul, we say, which departs for another world where it will join its forefathers. The body of the deceased is not regarded like a carcass of some animal: specific care must be given to it and a correct burial; not merely for reasons of hygiene but out of moral obligation. (2009 [1907]: 27)

Hertz's study of secondary burial ceremonies among the Dayak of Indonesia laid an important foundation for conceptualizing death as a *social* process in time, rather than simply a biological point in time. Hertz emphasized the mental and social processual aspects of mortuary rituals and analysed these as society's triumph over death: '(D)eath as a social phenomenon consists in a dual and painful process of mental disintegration and synthesis. It is only when this process is completed that society, its peace recovered, can triumph over death' (2009 [1907]: 86).

In another significant contribution to the study of death as a social phenomenon, anthropologists Maurice Bloch and Jonathan Parry (1982) developed and substantiated the argument that funeral rituals always include ideas about the regeneration of life through symbols of fertility and rebirth. In this book, we further explore ideas about social processes in death, not as 'societal triumph' over death, not as a mere method for studying the social per se, nor as a 'regeneration of life' (Bloch and Parry 1982), but as key to human understandings of temporality.

Our perspective for exploring questions of death and time is inter-disciplinary, with contributions from anthropology, cultural studies, media studies, Egyptology, sociology and the history of religion. All chapters take their point of departure in original empirical studies, rather than in philosophical and theological thinking.

The empirical cases are drawn from around the world and from different historical epochs: ancient Egypt, Denmark, Norway, the Orkney Islands, contemporary and historical Europe and the United States, Nepal, Melanesia, Jordan, India, the Philippines, Siberia, Nigeria and Uganda. Thematically, the chapters deal with a broad range of issues in relation to death and time: child death, reincarnation, reburials, headhunting, meat, witchcraft and divination, mummification, sainthood, landscape, mourning, pilgrimage, timing of death and longevity technologies. Through concrete empirical lenses, we explore the diverse relationships that humans have fashioned and continue to fashion between death and temporality.

Taming Death's Time

Events of death open and trigger cascades of questions about time, and potential answers and solutions have to be found. Death offers opportunities, we might say, to grapple with the abstract, existential and irreversible qualities of time in concrete and material ways. If there is a 'triumph' involved, as Hertz suggested, it has to do with our grasping and taming of time. We might, however, conceptualize the endeavour as less of a triumph and more as an act of appropriating specific qualities of time. The dead body offers up what Claude Levi-Strauss denoted 'the logic of the concrete' (1986) to comprehend, in an applied and tangible manner, the abstract qualities of time, whereby actions have to be taken to establish the dead person's entrance into the time of the deceased: corpses have to be prepared, graves have to be dug or fires have to be set, and so on. Rituals establish a temporality through actions with material objects in a set space. These social technologies help create 'the time of the dead' in an explicit, performed and material way.

We use the concept of social technology, inspired by philosopher Ian Hacking, who reminded us that technology means 'practical art' and pointed out that technology does not primarily concern abstract knowledge, but knowledge about how you 'do something' (1996: 80). In this volume we explore how people around the world apply various technologies – often and mostly social ones, such as ritual, but also including material technologies such as mummification and longevity procedures – to grasp time.

Death demands that the body has to be dealt with in what is considered as 'appropriate time' by society or a social group. By taking action on the dead body, we engage with ideas about time, such as eternity and repetition or duration and finality, through concrete timework. Death may be an occasion to face the hard fact that time simply passes, irrespective of any human interference. In fact, mortuary rituals often acknowledge this human powerlessness over time. At the same time, however, rituals may also be ways of demonstrating and exercising the human capability to govern at least some provinces of time. The kind of concrete material time invoked by the dead body on the living is transformed into an array of social time intervals, stages and rhythms; in fact, all death rituals comprise timing as a central structuring aspect (Leach 1961; Huntington and Metcalf 1979; Bloch and Parry 1982).

When death strikes our loved ones, most of us stand defenceless and have to face the irreversibility of time, what anthropologist Alfred Gell describes as 'objective time' (1992) and philosopher Paul Ricoeur as 'cosmological time' (1980). As human beings we have to cope with the fact that we cannot undo or reverse the flow of time. We cope by converting it into social and performative time (Flaherty 2011): forms of 'subjective' (Gell 1992) and 'narrative' time (Ricoeur 1980). By taming mortuary time with timework such as phasing, scheduling and sequencing in narrative and ritual, we attempt to grasp and take control over the relentless passing of time.

Events of death have strong potentials of evoking senses of time and questions about existence in time in a way that few other events have. When we are confronted with the corporeality of the dead body, we may be reminded not only of our own upcoming death, but also that fellow human beings will die. As human beings we grow up to learn about and live with the inevitable fact that we and our loved ones shall die some day. The tricky part is that the specifics – the concrete time and sequencing of our deaths – are unknown. We might assume that the old die first, but we also know that we cannot know for sure who will go first: who will have to deal with whom as dead? We do not know how we will die and when this will happen. It is getting closer every moment, we know it will come but are kept in suspense as to when. The uncontrollable, wild aspect of timing death is, we argue, reversed through various forms of timework in mortuary rituals as ways of taming, not death itself, as pointed out by historian Philippe Ariès (1982), but as a way of taming the unpredictability of timing.

The anthropologist Edmund Leach (1961), for one, has suggested that our inherently ambiguous concept of time facilitates the assertion that birth comes after death as day comes after night. The category of 'time', according to Leach, covers two quite different kinds of experience: time as repetitive and time as irreversible duration. He argues that by merging both kinds of time in the idea of reincarnation, we avoid recognizing life's irreversibility by assimilating it into repetition. Rather than reducing ideas about reincarnation as ways of coping with the finality of death, we wish, through this book, to make room for exploring practices and ideas as specific conceptions of time, ways of creating different forms of time and thus dealing with life issues in the broadest sense. Human beings tame time through all kinds of social technologies in order to gain a sense of control over life, and these actions may be triggered by death. Through such taming practices we acknowledge, on the one hand, that we are limited creatures, overwhelmed by the impersonal workings of objective time, and on the other that we can, to some extent, redefine time, life and the cosmos.

Death, as we profanely know it from the side of the living, means that a lifetime has ended. One body of tangible, pulsing and lived time has come to a definite end, or in some conceptions, to a point of potential transformation, turning into something else. An actual, physical death may leave a sense of non-time with the mourners, because a body of time has been taken out of their world in the sense that their time together with this person has been discontinued.

Death as Event and Relation

There is an important sense in which it is the awareness of our upcoming death that makes us the beings that we are. Most human beings differentiate themselves from the divine by their mortality (at least according to many religions) and we usually regard ourselves as distinguished from other animals by our experience as mortals. Our consciousness of death – inseparable as it is from our awareness of the passing of time – is the essential distinguishing mark of what it is to be human. How, then, does death as the end of 'my time' shape our existence? According to Martin Heidegger, who is often taken to exemplify the philosophical preoccupation with death and its relationship to time, the answer is that

> Death does not just 'belong' to one's own Dasein [Being-there] in an undifferentiated way; death *lays claim* to it as an *individual* Dasein. The non-relational character of death … individualizes Dasein down to itself. (Heidegger 1962: 306–8, emphasis in the original)

What Heidegger seems to suggest in this far from self-explanatory prose – and what we set out to question – is that in death, my social relationships to others are annulled, and I experience myself as 'authentic', my own person, as opposed to the anonymous identity given to me by the crowd. Death in this sense signifies an individual, spiritual liberation and it is for this reason that Heidegger argues that our lonely finitude is not to be resented and avoided as it brings 'an unshakable joy' (Heidegger 1962: 358).

To some philosophers of the Heideggerian tradition, death might be considered to be the ultimate event per se – singular, irreversible and unfathomable (see, for example, Derrida 2003 and Badiou 2005). However, we take our inspiration from the French philosopher Claude Romano, who offers keen reflections on the eventual and relational character of death (Romano 2009) that can qualify as a theoretical departure from Heidegger. Romano asserts that 'an original dimension of the human adventure is a constitutive relation with others … [not] as in Heidegger's *Mitsein* [Being-with], but in the sense that [to the individual] others always become evident as such … starting from events' (2009: 114).

The death of another, his or her 'loss of the world', as Romano puts it, cannot in itself be an experience to me, since it is the other and not me who dies. However, three experiences are accessible to the bereaved: firstly, the experience that death as an event in itself is nothing, and that this nothing *is* death; secondly, the curious experience that death as an event can touch one's heart so deeply, without appearing as a visible entity in the world; and thirdly, the fact that the dead disappear, which may produce a cataclysm through which the one who has lost someone loses her or himself – that is, a part of the living person dies along with the person dying. It is in this fashion that we are intertwined with the dead in a shared social reality. People learn how to become themselves again without the person who is gone. People handle the breakdown of their worlds, their cosmos and time, in order to re-create themselves in the world and with others.

Whereas Romano, like most other Western philosophers, focuses on death as an endpoint and nothingness, we emphasize the notion of death as a continuation of sociality, though in altered ways. It is from this standpoint – of death as/being an integral part of social life – that this book challenges Heidegger's belief in an authentic being-towards-death as superior to life as part of the faceless masses. We venture to suggest that quite the opposite is true: death is inherently social in the sense that it pre-exists us, we handle it socially and we experience death through the death of others. It is always already in place, waiting to assign us a place within its sphere and to/thereby redefine our social relations. It is often not understood as finitude per se, but as a transitory realm that transports us into another state of communal existence.

Thoughts around death as an all-inclusive whole of undifferentiated existence are found in psychoanalyst Sigmund Freud's famous conception of the 'death drive', the notion that all animate life is destined to cancel itself out (1991: 38). Freud's theory of the death drive came about as a result of his observations on how some mourning turns into pathological melancholia (Freud 1917). It was this fact that made Freud put forward a new theory based on the fundamental conflict between the life and death drives. The death drive, according to Freud, is the most fundamental drive, so that the imperceptible drive to die inevitably lies behind our more evident drive to live. Several of the book's contributions in one way or another elaborate Freud's audacious thesis that the fundamental terms which constitute human existence are the struggle between life and death, between potency and powerlessness which, rather than being opposites, are contained within each other.

Timework: Doing Death's Time

Studying the cases in this volume, one issue in particular is obvious: time is *not* a given and neutral phenomenon. It is moulded, grasped and transformed in various ways. Sociologist Michael G. Flaherty (2011) has introduced the concept of 'timework', which basically refers to what we have labelled 'social technologies of time' through which humans preclude, craft and produce certain temporal experiences. Flaherty makes a distinction between different analytical categories of time: duration, frequency, sequencing, timing, allocation and taking time (2011: 12) that are illuminating in analysing death practices.

Given the intrinsic connections between life, death and time that become apparent during the crises of death, it is not surprising that when humans face death, timework is one of a number of performative practices engaged in. While many of these practices might analytically be considered timework, in practice they are often regarded to be about other things by the performers. It is noteworthy that timework is not necessarily a conscious or reflective activity, but might have the character of 'timescapes', that is, actions practised in an immediate and unreflected way (Bluedorn and Standifer 2006). We would argue, however, that in relation to death, questions of time and the timework involved become more pronounced than in most other time-practices. Death 'stops us', at least for a while, and forces us to deal with practical as well as more existential questions concerning time. When a death has occurred, we cannot simply go on with everyday issues and rhythms as if nothing has happened.

Several of the accounts in this book promote the idea of a past that does not cease to be once the present takes over, but rather a past that could never really exist if it did not coexist with the present of which it is the past. This in turn suggests a rather widespread cross-cultural way of thinking about time, as a kind folding into each other of past, present and future; not as fully identical, but with an important element of endurance and repetition.

Henri Bergson (2004 [1912]) and his philosophical protégée Gilles Deleuze (1994 [1968]) offered another route to a somewhat similar conclusion, when they problematized a whole range of assumptions regarding our conceptions of the present and the past. The past, they argued, is not simply time gone by. Rather, the past lives a shadowy existence and exerts its influence through its capacity to link up with and thus inform the present. For this reason, the present is not purely something 'in itself', but includes both past and present. A 'new' present could never replace an 'old' one if the latter did not pass while it is still present. Thus, Bergson and Deleuze contested our commonsense view of the succession of the past by the present, by postulating the simultaneity of past and present.

Ritualizing Death and Time

Time is inherent in ritualizations in the sense that in rituals we 'make time' and that time creates us through the rituals. Rituals help us sequence events, by ascribing the proper duration to our actions; as such, rituals are a way of simply coming to grips with time. Our take on ritual does not only consider timework as a healing method, but also as a way of re-appropriating time.

Taking our cue from Hertz (2009 [1907]), who pointed to three aspects to be considered when analysing death – body, soul and mourners – we remark how these three foci are also useful for understanding time in relation to mortuary rituals: (1) time related to the body (e.g. finality, decomposition or longevity); (2) time related to the soul (e.g. eternity, repetition and continuity); (3) time related to the mourners (e.g. re-creation, maintenance and renewal of the social). Most of the contributions in this volume foreground one of these aspects (such as the time of the soul), while also considering the other kinds of time that are handled simultaneously (such as the time of the body and the mourners). This multi-layered timework is an important aspect of ritualization.

Catherine Bell (1997: 168) has defined ritualization as 'the simple imperative to *do* something in such a way that the doing itself gives the acts a special or privileged status'. This is in line with Caroline Humphrey and James Laidlaw (1994), Inger Sjørslev (2007) and Adam B. Seligman et al. (2008), who all agree that ritualization is a certain framing of actions, the taking on of a certain mode, within which actions are performed, and that it is this framing or 'ritual attitude' which is the crucial point; not the action itself. The meaning we read into ritual often comes into play outside the frame of the ritual itself (Seligman et al. 2008: 4). Meaning might be unfolded by ritual participants in elaborate explanations with a reference to tradition or participants might refuse to give any hints with regard to the meaning by simply referring to the tradition of the ritual itself (Seligman et al. 2008: 4). Ritual, therefore, is not primarily a referent of meaning, but perhaps more a negotiation of meaning or a subjunctive experimentation with what meaning is and how it is practised and socially distributed and shared. More importantly for our argument, though, ritual is time making.

While it is impossible for the mourners to experience death in the phenomenological sense, as Romano argues, we claim that one way of actually experiencing death is through its ritualizations. Rituals can, on the one hand, create experiences of clear-cut differentiations, such as that between the sacred and the profane (Smith 1987) while, on the other, allowing for meta-reflections on its framings, making room for endless negotiations of meaning (Bateson 2000 [1972]). Seligman suggests that one of the crucial properties of ritual is the experimental 'as if', creating a shared subjunctive world (Seligman et al. 2008). Taking this idea further, we argue that doing time is a crucial part of this shared subjunctiveness.

Rites of passage in the sense described by Arnold Van Gennep (1961 [1909]) establish periods, such as childhood and adulthood, and without these ritualizations the distinctions between these periods would not exist. In this sense, the temporality of the ritual is in fact a kind of 'calculated action', a technology of time manipulation, in that the delay or time lag generated between the successive stages of the ritual is what allows human social life to evolve. Time, therefore, is not something we just measure as if it existed as a thing in itself. Rather, as Leach pointed out, 'we *create time* by creating intervals in social life' (1961: 135, emphasis in the original). Timework, such as in rituals, is a means to this end: 'Until we have done this there is no time to be measured' (Leach 1961: 135).

The Agency of the Dead

If indeed the dead are integral to the social matrix in which they take an active part, does this then imply that we can analyse the dead within the long anthropological tradition of focusing on the 'social agent'? (see, for example, Barth 1993; Bourdieu 1977). Alfred Gell encapsulates the term 'agency' comprehensively when he writes:

> Whereas chains of physical/material cause-and-effect consist of 'happenings' which can be explained by physical laws … agents initiate 'action' which are 'caused' by themselves, by their intentions. (1998: 16)

Since its beginning, anthropology has been preoccupied with a series of problems to do with the apparently peculiar relations between human beings and objects, plants, animals and other types of beings that somehow appear as, or do duty as, social agents. This problem was initially announced by Edward B. Tylor (1958), where he discusses animism as the 'primitive' inclination to 'impute a soul [that is, social agency] to that which empirically does not have one' (cited in Stringer 1999: 552). However, one of the problems with Tylor's analysis – which has somehow shadowed the field ever since – was his failure to acknowledge a fundamental difference between the social agency ascribed by so-called 'savage' peoples to animals, plants and things, and the agency which the people acknowledge is an inherent part of human social life (Willerslev 2007: 21). From the case studies of this book, it is quite clear that the social agency of the dead differs from that of the living in important ways and that we cannot simply collapse one type of agency into the other. While living human beings form what are evidently social relations with dead people, such relationships are not equivalent to those formed among the living themselves, just as the agency exercised by the dead is not equivalent to that exercised by ordinary living human beings. What then are the differences? The ways in which social agency can be invested in the dead, or can emanate from the dead, are exceedingly diverse. The dead are often considered as being more potent than the living and being able to do extraordinary things.

The ability of the dead to act on the living is not so much tied to questions of intentionality, what the dead choose to do, as to potentiality, what they could do. Inferred in this is the important insight of Gell in relation to the agency of objects, but which is equally true of the dead; namely, that 'we recognize agency, ex post facto … that is we cannot tell that someone is an agent before they act as an agent, before they disturb the causal milieu in such a way as can only be attributed to their agency' (Gell 1998: 20). In other words, the dead are not necessarily considered agents *ex ante*, but they have the potentiality to act as agents and, when they do, they can cause extraordinary events to happen in their vicinity. However, the agency of the dead always has to be considered by the living in the subjunctive mode (Whyte 1997), because we cannot know for sure whether something happening is related to the dead or not, just like we cannot know what will happen if we do not perform certain rituals. Would time stop? Probably not, but we may as well do them. This is perhaps why rituals need to be conducted – as a kind but firm way of reminding time that the living are in charge.

The relationship between the living and the dead is, furthermore, marked by fundamental aspects of asymmetry and hierarchy. The deceased are often assigned a permanently superior status vis-à-vis the living because they are considered as residing in closer proximity to the divine or because their world, although at times considered to be an exact replica of this world, does not have its troubles and sins (Willerslev 2009). As Edmund Leach (1976: 81–93) pointed out several years ago, in many societies, the world of the dead is generated by direct inversion of the characteristics of ordinary experience: 'This world is inhabited by mortal, impotent, men [*sic*], who live out their lives in normal time in which events happen in sequence, one after another. In this world we get older and older "all the time".' The world of the dead, however, is often inhabited by immortal, omnipotent ancestors, 'who exist perceptually in abnormal time in which past, present and future all coexist simultaneously … The souls of dead men pass from the normality of This World … to become immortal ancestor deities in the Other World' (Leach 1976: 82–3).

In some cases divinity itself is not accessible to humans, but is rather like the Creative Spirit of the Nuer, whom E.E. Evans-Pritchard (1956: 4–7) described as being beyond contact and comprehension, yet the giver and sustainer of all life. In these cases, there is a real sense in which divinity is synonymous with time itself, the changing seasons and the duration of life and death, which lie beyond human control. Still, among the various functions which the technologies of ritual may fulfil, one that is very important, which is documented by the various case studies in this book, is the ordering of time.

The cases of this book show how death is perhaps the event par excellence for unfolding how people around the world create and re-create time through various social technologies. Death does not stop time but creates time socially and provides access to other possible realms of time.

The Contributions

The chapters of this book are not meant to form an all-conclusive argument built up layer upon layer. Instead, they should be seen as diverse disciplinary facets of the same basic enigma: the relationship between death as a material phenomenon and human conceptions of time. Each chapter presents a distinct empirical, disciplinary and theoretical perspective on the enigma. Yet, together they form a kind of mosaic of perspectives, progressively adding different light to how the enigma might be addressed and to its possible explanation.

The chapters are grouped under four headings: (1) conceptualizations of death, materiality and time; (2) death's time: mourning and remembrance; (3) living with the dead: ancestors, landscape and agency; (4) self-death: corporeality and willed death. We describe them here in turn.

Conceptualizations of Death, Materiality and Time

Death is as much part of our social existence as is life. Our 'being-towards-death' is not simply about an individual, existential, private region 'inside' us, but is essentially an effect of our relations with others. Death is, so to speak, 'between' us rather than 'within' us. Our death-to-come is, therefore, not reducible to the human loneliness and finitude of Heidegger's philosophy, but involves a vast, tangled network of social relations, including both the living and the dead, which weaves itself through us into the very core of our being. In the book's first three chapters, we find examples of this very matrix of social relatedness that is organized and sustained in and through death.

Stuart McClean's study of the Orkney Islands in Chapter 1 suggests that death needs to be considered in relation not simply to human beings but also to the islands' spectacular nature – a perspective that includes environmental, ecological and geological topographies. McClean uses an idea of the 'invisible crowd' to argue that the dead are an 'ocean', an undifferentiated mass, which is ontologically prior to any conceptions of the unique, named and personalized dead. That is, the individual dead, whose names are imprinted in obituaries and inscribed on headstones, are initially animated, perceived and remembered out of the formless mass, from whom they originally derive their personified existence and to whom they return when fading out of memory. In this way, McClean argues, life cannot be defined as individual 'being-towards-death', but rather precedes us as an anonymous collective realm that is already in place, waiting to ascribe us within it.

This idea of death as closely linked with a mass or anonymity is also prominent in Knut Rio's description of divination and witchcraft in the Melanesian nation of Vanuatu, in Chapter 2. Rio argues that as forms of social agency, both divination and witchcraft are in fact 'technologies of death' in the sense that they draw upon a prior or more fundamental realm of 'non-being'. Divination aims at looking into the past to discover the cause of death and does so through the manipulation of simple objects: stones, shells, bones or thorns. The crude simplicity of the objects is significant in that diviners operate towards a realm of nothingness by using objects that are 'next to nothing'. As a social force the reality of witchcraft is present by absence, it is potential and intangible, and can hardly be foreseen, just like death. Divination as a technology of time makes use of the agency of the dead and the spirits to see into other realms of being and time. Rio explores how divination works to recall past moments in time that are related to the potent moment of the present, when someone dies or falls ill from witchcraft or sorcery. Divination seeks into the space of death by manipulating material objects to speak about the space beyond life and materiality, beyond the realm of 'being-here and being-now'.

In Chapter 3, Michael Kearl and Michael Hviid Jacobsen argue that human awareness of mortality has shifted, in late modernity, from a focus on the religious afterlife to late life and its end, which reflects a new concern with life *before* death, rather than *after* death. Kearl and Jacobsen argue that while scientism and religious fundamentalism are two distinct 'death systems', in reality they are often combined in death practices. The chapter analyses various contemporary public representations of death in late modernity: from an older person's gradual death and images of death in (American) popular culture, to science's war against death. They point to cremation as an acceleration of the dissolution of corporeal selves, and to the spread of 'bucket lists' as exemplifying current practices, and argue that this tendency to focus on life before death is a new kind of preoccupation with eternity.

Death's Time: Mourning and Remembrance

Mourning and remembrance is essential to death's sociality. Both are evidence of the fact that social ties exist and continue to exist between the living and the dead.

However, for Freud and his influential theory of psychoanalysis, in order to reach a healthy resolution the mourning process needs to come to an end: the bereaved must detach themselves from the deceased and accept that acts of love can no longer be returned. But is this truly how people around the world go about dealing with their losses? Not if we are to believe the three contributions here, exploring the practices of mourning in places as diverse as Siberia, Denmark and India.

Rane Willerslev, in Chapter 4, discusses Freud's death drive, and revitalizes it by coupling it with Deleuze's philosophy of time and Siberian beliefs in rebirth. Freud saw repetition as an indispensable key to unlocking the mystery of the death drive, but he was incapable of conceiving the human compulsion to repeat as anything other than the repetition of the same. Freud's narrow understanding of repetition, Willerslev suggests, is the real reason why he saw the mourner's inclination to invest his libido in the deceased as the direct route to melancholic depression. Willerslev shows that in the Siberian north, the rebirth of ancestral souls into newborn children is not simply repetitious of the same, but also of the different. Rebirth consists in the return, not of the identical, but of someone other. This allows the mourner to repeat her emotional investment in the dead without regressing into melancholia, exactly because repetition always leads to the return of something other than what was, something new. Willerslev proposes a new reading of Freud's death drive as an uncanny excess of life that persists beyond the biological cycle of life and death and which finds its creative equivalents in Deleuze's concept of the 'virtual' and the Siberian peoples' notion of the 'soul'.

In Chapter 5, Dorthe Refslund Christensen and Kjetil Sandvik argue that the death of a child in Denmark does not simply imply the end of a baby's short life. It also ends the imagined future of the parents, who would have planned and envisioned various phases of life to be shared with their child. The ending of the child's life, then, causes the parents' time horizon to collapse and their world to radically change. On the website Mindet.dk, the parents get to share these experiences with other bereaved parents. This is not only about the parents keeping an ongoing relation to the dead child. Through different narratives and ritualizations the parents perform various kinds of timework, thus re-creating their sense of differentiated time and their ability to enrol themselves back into everyday social life. Drawing on the narrative theory of Paul Ricouer, the chapter analyses how life, or rather the present, after the meaningless and devastating death of the infant, is re-established as a meaning-making process by ways of re-working and re-telling the past, online.

Piers Vitebsky, in Chapter 6, describes the Sora of tribal India, who engage in elaborate dialogues with their dead. The dead are initially deemed to be in limbo, separated from, yet still close to, their domestic environments. They do not easily go to their lineage ancestors but are pulled back by the circumstances of their individual deaths and the tangle of their personal relationships. This is considered highly dangerous as all illness and death is believed to be caused by dead people who have suffered from similar misfortunes. For this reason, the living need to maintain prolonged dialogues with the dead ghost-souls through a shaman intermediary, who must persuade the dead not to harm their living kin. However, this elaborate system in which the living and the dead gradually set each other at ease is rapidly disappearing. With their conversion to Christianity the Sora are left in a kind of existential limbo in relation to the dead, with whom they no longer have any contact. Vitebsky describes the deep-felt anxiety among elderly Sora, who are fearful of their solitude after dying as there will be no one left to talk to them.

Living with the Dead: Ancestors, Landscape and Agency

The following four chapters deal with how we live with the dead as ancestors, in the landscapes or in other forms of materialized agency. They clearly point to the fact that often the dead are not conceived of as something purely spiritual, ghostly or incorporeal. Rather, they have a tangible reality in the form of physical imprints, graves, tombs and, not least, through the concrete effects they have on the life of the living.

In Chapter 7, Mikkel Bille describes how the Bedouins of Jordan are surrounded by the graves of ancestors, some of whom have turned into saints with the capacity to influence the lives of the living beyond the range of normal human action. Despite the fact that a more scriptural understanding of Islam is spreading in the area, which rejects the sainthood worship as belonging to 'times of ignorance', the agency of the graves keeps manifesting itself in the lives of both locals and pilgrims in the most conspicuous ways, including curing illnesses and safeguarding the rains and fertility. In this Islamic context, these extraordinary post-death events are taken as a sign of the dead person's closeness to God, so that the deceased effectively mediate between the divine and earthly worlds. Bille enters into an elaborate discussion of how we might define the agency of the dead and he points to key features that have bearings far beyond his own region of study. Thus, Bille unfolds the widely reported ethnographic fact that often the agency of the dead is understood in terms of an 'excessive agency' – that is, some persons may be so full of agency that it continues, even expands, after death beyond the normal range of possible human action.

Jesper Østergaard, in Chapter 8, also discusses the agency of the dead, here in relation to how Buddhist pilgrims negotiate religious ideas about immortality, long life, and their own individual death through ritualized interactions with the landscape at the Buddhist and Hindu pilgrimage site of Maratika in Nepal. According to local Buddhist tradition, the religious figure Padmasambhava achieved his immortality in mythic time at this site by surmounting a demoness and removing a range of obstacles. The pilgrims find themselves able to read concrete physical imprints of these events in the actual landscape. They consider signs in a cave to be the footprints of Padmasambhava and so the whole landscape is regarded as blessed and thus bestows longevity. Through the performance of various site-specific rituals that are rooted in the material and in the body, such as eating the dirt and squeezing their bodies through narrow passages in the physical landscape, Buddhist pilgrims enrol themselves in the karmic circle inherent in reincarnation. In these pilgrim practices in the landscape that is saturated with the agencies of mythic figures and periods, different time scales, stretching from individual to cosmic, are effectively connected.

Lotte Meinert and Susan Reynolds Whyte describe, in Chapter 9, how the Acholi in northern Uganda use re-location of the dead in the landscape to influence the future. In the process of moving back to their rural villages after more than two decades of conflict, people exhume and rebury their dead, who were buried wrongly. This is considered a social technology for creating and re-creating specific pasts, presents and futures. The concreteness of the dead body offers possibilities for manipulating time at multiple scales. The re-interment of bones in their proper homes marks, not only the transition to the status of spirit, ancestor and past, but also marks the re-establishment of a properly organized home in the present with concrete claims to a specific piece of land for future generations. In Acholi cosmology the reburial of the bones in the ancestral land creates material continuity between land and people, as well as between the living and the dead, which makes new life possible. The historical situation of post-conflict reburials in northern Uganda provides a window, not only for understanding the timework carried out in this situation, but also for exploring how the materiality of dead bodies – in general – can be used and analysed as a technology of time manipulation at different time scales, simultaneously.

In ancient Egypt, described by Rune Nyord in Chapter 10, biological death is only the first step in a social transfiguration into permanence of which the mummification process is the tangible index. Mummification as a social technology of time manipulation thus transforms a lived past into an ideal past that lasts into eternity. This transformation corresponds to what Nyord denotes the deceased's ideal biography, which is also inscribed in hieroglyphs on the tomb. The text casts the dead as an exemplification of a prototype of virtuous behaviour, which unites his supreme career with those of other tomb owners of an equivalent social status. Drawing on Deleuze's philosophy of time, Nyord argues that the ideal biography signifies the 'pure past'. Unlike the past dimension of the present, the pure (or virtual) past has never been present, but instead acts as the 'ground' of time per se, that causes the present to pass. Nyord describes how the Egyptian tomb can be seen as a lens capturing and refracting the past, present and future in various specific ways and how this way of living with the dead structured the entire Egyptian cult.

Self-death: Corporeality and Willed Death

The last four chapters in the book deal with various forms of self-death, including confrontations with the corporeality of death and actions of willed death. We see our own death in the death of others, including other beings. This may be a fearful reminder of one's own death or it can be deliberately used to kill off aspects of oneself that are no longer desirable. The will to self-death may also be exercised by attempting to control the exact timing of one's death.

In Chapter 11, Kristian Bjørkdahl and Karen Syse describe how photographs of dead animals have been gradually removed from Norwegian cookbooks over the last decades. The first edition from 1955 is quite explicit in exposing dead animal bodies, including recipes for how various parts of an animal can be used in food. In the latest edition from 2002, however, virtually all these links between animals and meat have been removed. The reason is the uncomfortable fact that dead animals remind us of our own upcoming death with all this entails of tangible flesh and bones. Every age and every culture use distinctive technologies of symbolic transformation that makes the killing and consuming of animals acceptable. In Lowland Amazonia, for example, a special trap is constructed that turns fish into vegetables, which allows the fish to be eaten without any feelings of guilt. However, in Scandinavian, (post-)modern society, there is no effective way to justify the transformation of animals into meat, none other than to hide and conceal it.

Henrik Hvenegaard Mikkelsen, in Chapter 12, reports on headhunting as a rite of passage among the Bugkalot in the Northern Philippines. Among the Bugkalot, a male child is regarded as a disgraceful being who, through confrontations with his own ignorance and incapacity to act independently of others, is exposed to public shame. However, by killing and taking the head of a random person outside the community and ridiculing the head by tossing it around, the Bugkalot young man engages in a kind of deliberate 'self-death' whereby he cuts off a socially dependent part of himself. This transforms him into a full male member of his community. This, however, is not an inner existential condition that is constitutive of our very being, but a ritualistic passage in time; a provisional step towards manhood in the eyes of others. Moreover, the fact that it is a person's head or rather his *face* that is removed, is of deep existential importance to the event. The face, according to Emmanuel Levinas (1969 [1961]) on whose work Hvenegaard Mikkelsen draws, enforces an ethical relationship, in that it more than any other part of the human body is related to our capacity for empathy. For this reason the Bugkalot themselves regard headhunting with a good deal of moral ambivalence; it is something admirable, even necessary, yet it is also an anti-social and ethically dubious act.

The taming of the time of death takes a very concrete turn in Chapter 13 by Murray Last. Based on many years of fieldwork and burial registration in the Muslim part of northern Nigeria, Last documents a significant over-representation of mortality on Fridays. Based on this material, anecdotal evidence, religious traditions and 'ethnothanatologies' the chapter examines the question: how far does the person dying have agency in the timing of their own death? It is suggested that the dying person actually has some control over the exact timing of his or her death despite cultural taboos that deny it. Last describes how the dying and their relatives actively attempt at influencing the timing of death to happen on a Friday, the holy time of the week for Muslims, and many do in fact succeed with this. However, they do not conceptualize and talk about the human agency in timing death. In their view the timing of death is entirely in the hands of God.

References

Ariès, P. 1982. *The Hour of Our Death*. New York: Vintage Books.

Badiou, A. 2005. *Being and Event*. New York: Continuum International Publishing Group Ltd.

Barth, Frederik. 1993. *Balinese Worlds*. Chicago: University of Chicago Press.

Bateson, G. 2000 [1972]. *Steps to an Ecology of Mind*. Chicago: University of Chicago Press.

Bell, C. 1997. *Ritual Perspectives and Dimensions*. New York: Oxford University Press.

Bergson, Henri. 2004 [1912]. *Matter and Memory*. Dover: Philosophical Classics, Dover Publications Inc.

Bloch, M. and Parry, J. 1982. *Death and the Regeneration of Life*. Cambridge: Cambridge University Press.

Bluedorn, A.C. and Standifer, R.L. 2006. Time and the Temporal Imagination. *Academy of Management Learning and Education*.

Bourdieu, Pierre. 1977. *Outline of a Theory of Practice*. Cambridge: Cambridge University Press.

Deleuze, G. 1994. *Difference and Repetition*. Trans. P. Patton. New York: Colombia University Press.

Derrida, J. 2003. A Certain Impossible Possibility of Saying the Event. *Critical Inquiry*, 33, 441–61.

Evans-Pritchard, E.E. 1956. *Nuer Religion*. Oxford: Oxford University Press.

Flaherty, M.G. 2011. *The Textures of Time: Agency and Temporal Experience*. Philadelphia: Temple University Press.

Freud, Sigmund. 1917. Mourning and Melancholia. *The Standard Edition of the Complete Psychological Works of Sigmund Freud, Volume XIV (1914–1916): On the History of the Psycho-Analytic Movement, Papers on Metapsychology and Other Works*, London: The Hogarth Press and the Institute of Psycho-analysis, pp. 237–58.

Freud, Sigmund. 1991. *On Metapsychology – The Theory of Psychoanalysis: 'Beyond the Pleasure Principle', 'Ego and the Id' and Other Works*. Penguin Books Ltd.

Gell, Alfred. 1992. *The Anthropology of Time: Cultural Constructions of Temporal Maps and Images*. Oxford: Berg.

Gell, Alfred. 1998. *Art and Agency: An Anthropological Theory*. Oxford: Clarendon Press.

Hacking, Ian. 1996. Memory Science. Memory Politics, in *Tense Past: Cultural Essays in Trauma and Memory*, edited by P. Antze and M. Lambek. New York: Routledge, pp. 67–88.

Heidegger, Martin. 1962. *Being and Time*. New York: Harper & Row.

Hertz, Robert. 2009 [1907]. *Death and the Right Hand*. Aberdeen: Cohen and West.

Humphrey, Caroline and Laidlaw, James. 1994. *The Archetypal Actions of Ritual: A Theory of Ritual Illustrated by the Jain Rite of Worship*. New York: Oxford University Press.
Huntington, R. and Metcalf, P. 1979. *Celebrations of Death: The Anthropology of Mortuary Ritual*. Cambridge: Cambridge University Press.
Leach, E.R. 1961. Two Essays Concerning the Symbolic Representation of Time, in *Rethinking Anthropology*. London: Athlone Press, pp. 124–36.
Leach, Edmund. 1976. *Culture and Communication: The Logic by which Symbols are Connected. An Introduction to the Use of Structuralist Analysis in Social Anthropology*. Cambridge: Cambridge University Press.
Levinas, Emmanuel. 1969 [1961]. *Totality and Infinity: An Essay on Exteriority*. Trans. Alphonso Lingis. Pittsburgh: Duquesne University Press.
Levi-Strauss, C. 1986 [1962]. *The Savage Mind*. Chicago: University of Chicago Press.
Ricoeur, P. 1980. Narrative Time. *Critical Inquiry*, 7(1), 169–90.
Romano, C. 2009. *Event and World*. New York: Fordham University Press.
Seligman, A.B., Weller, R.P., Puett, M.J. and Simon, B. 2008. *Ritual and its Consequences: An Essay on the Limits of Sincerity*. New York: Oxford University Press.
Sjørslev, I. 2007. Ritual, Performance og Socialitet: En Introduktion, in *Scener for Samvær: Ritualer, performance og socialitet*, edited by I. Sjørslev. Arhus: Aarhus University Press, pp. 9–34.
Smith, J.Z. 1987. *To Take Place. Toward a Theory in Ritual*. Chicago and London: University of Chicago Press.
Stringer, M.D. 1999. Rethinking Animism: Thoughts from the Infancy of Our Discipline. *Journal of the Royal Anthropological Institute*, 5, 541–56.
Tylor, Edward B. 1958 [1871]. *Primitive Culture: Researches into the Development of Mythology, Philosophy, Religion, Language, Art, and Custom*, vol. 1. London: John Murray.
Van Gennep, Arnold. 1961 [1909]. *The Rites of Passage*. Chicago: University of Chicago Press.
Whyte, Susan Reynolds. 1997. *Questioning Misfortune: The Pragmatics of Uncertainty in Eastern Uganda*. Cambridge: Cambridge University Press.
Willerslev, Rane. 2007. *Soul Hunters: Hunting, Animism, and Personhood among the Siberian Yukaghirs*. Berkeley: University of California Press.
Willerslev, Rane. 2009. The Optimal Sacrifice: A Study of Voluntary Death among the Siberian Chukchi. *American Ethnologist*, 36(4), 693–704.

Chapter 1
"Seaweeds and Limpets will Grow on Our Grave Stones": On Islands, Time, Death and Inhuman Materialities

Stuart McLean

"Sorry for your loss." The phrase is a staple of mass-produced bereavement cards and the rhetoric of professional funeral directors (think *Six Feet Under*, for example). Yet, generic as it seems, it speaks always of the loss of a particular someone—a family member, friend or loved one. Moreover, it is, invariably, a phrase spoken by the living to the living, offering comfort and consolation in the name of one who was once but is no longer among them. But how might the dead themselves speak of loss if they were able to do so? Might they speak to us not of individual bereavement but of a loss that is absolute, all-consuming and without recompense but that is, nonetheless, the indispensable precondition for … well, everything? To imagine what the dead might tell us about loss though, we need to think of them not as the particularized individuals whose names are printed in obituaries or inscribed on headstones but as an altogether more diffuse and expansive presence, one that infinitely exceeds that of the living. Such a presence, I propose, can be grasped most effectively not by examining beliefs and practices specific to a particular cultural-historical context (for example representations of the dead as individuated ancestors or as an unquantifiable mass) but by juxtaposing accounts from a variety of times and places, including my own current research on relations between the living and the dead and perceptions of long-term environmental change in the Orkney Islands, off the northernmost tip of Scotland. The spatiotemporal shifts that follow are intended less to establish links of comparison across periods and localities than to effect an evocative conjuration of time and matter as the impersonal ontological basis for the making and unmaking of human worlds. It is in these terms, I shall suggest, that the expressions of the living regarding the dead—verbal, visual, performative, architectonic—can be most suggestively understood.

A further wager of this chapter is that islands might have something distinctive to teach us about the ways in which time and matter are implicated both in relations between the living and the dead and in the shaping of humanly significant cultural worlds. Anthropologists, of course, have often written about islands, whether as the paradigmatic setting for Malinowskian participant-observation fieldwork or as spaces of intercultural contact and exchange—for example between Europeans and the various populations they encountered through voyages of trade, exploration or colonial expansion (Dening 1980; Malinowski 1984; Sahlins 1985). Drawing on my own research in Orkney, along with a range of archival and documentary sources, I aim here to consider the material specificity of islands as settings for a variety of encounters of a rather different kind: between the living and the dead, between solid and liquid modalities of matter, between humans and a range of other-than-human materialities and, not least, between human-centered, culturally calibrated time (including both historical time and the more expansively conceived temporalities of environmental and geological change) and the more definitionally elusive, impersonal and non-linear time of becoming and dissolution from which, I suggest, the latter must be understood to derive. Given its geographical location and its long history of human settlement (stretching back perhaps as much as 9,000 years), Orkney is at once a landscape densely marked with the enduring physical after-traces of the dead of past generations and one subject to continuous refashioning through the ongoing marine erosion that will, eventually, cause the islands to vanish beneath the waters of the North Sea from which they first emerged (Finlayson and Edwards 2003: 122–3; Wickham-Jones 1994: 59–63, 74).[1] Articulating an account of the disposal and memorialization of the dead in Orkney with an overview of long-term environmental change and its associated imaginaries, I develop an understanding of death, death ritual and mortuary practices less as expressive of social relations among the living or of culturally specific beliefs of the living about the dead than as a generative interface between a self-designated human realm and the processes of material transformation and dissolution by which it is at once constituted and continuously threatened.

Islands before and after History

The short essay "Desert Islands" (1953) is an early work by the philosopher Gilles Deleuze. Deleuze was interested in islands primarily as a source of creativity, of new beginnings, demanding as such a particular kind of attunement to one's environment on part of their human settlers. According to Deleuze there were two kinds of islands—"continental" ones, broken off shards of existing landmasses—

1 The material evidence of life in the Mesolithic period (c. 10000–5000 BCE) in Orkney is scant, consisting principally of a number of "lithic scatter" sites (at Seatter, South Ettit, Wideford Hill, Valdegar and the Loch of Stenness), which have yielded small polished stone tools and chippings. A charred hazelnut shell, recovered during excavations at Longhowe in Tankerness in 2007 has been dated to 6820–6860 BCE. A recently excavated site on the northerly island of Stronsay has produced pieces of flint with a tentative dating of 7000 BCE or older and may prove to be the oldest settlement in Orkney. There is, however, no evidence to suggest that these sites were in year-round occupation and no Mesolithic tombs have been found in Orkney, or, indeed, anywhere in present-day Scotland (Finlayson and Edwards 2003; Towrie 2008; Wickham-Jones 1994: 59–74, 2007).

and "oceanic" ones—the result of the petrification of coral reefs or of magmic upwellings from beneath the ocean floor, rising, cooling and solidifying to create new formations. Both kinds of islands testified that the elemental "strife" of land and sea was not yet at an end (much as humans residing in more populous localities might like to pretend otherwise). They thus served as a reminder that nothing was settled, that the forms of the existing world were not definitively fixed. The strife of land and water was one that preceded and would continue beyond the limited time-span of humans' presence on earth: "Islands are either from before or for after humankind" (Deleuze 2004: 9). In this sense, for Deleuze, all islands were, in effect, desert islands, belonging to a time at once pre- and post-historical. Human imaginings about islands participated in and carried forward the movement out of which islands themselves emerged. The arrival of humans, whether as refugees, colonists or castaways, did not put an end to the desertedness of desert islands but rather completed it, bringing it to consciousness of itself. But the individual human imagination, according to Deleuze, was not capable on its own of immersing itself fully in the movement—the *élan*—giving rise to islands. To do so required the engagement of the collective imagination, as revealed most profoundly in the guise of ritual and mythology. Crucially, for Deleuze, mythology was not an explanatory framework imposed upon reality, not something simply "willed" into existence, but a human creative act participating in and giving expression to the other-than-human creative energies and forces that produced desert islands themselves as privileged sites of new beginnings (Deleuze 2004: 10–12). Deleuze, however, sees the creativity exemplified by islands is as a matter of re-creation, of beginning again rather than beginning from scratch. The point is an important one because the second creation, the act of re-creation, affirms creativity itself as an ongoing and endlessly open-ended process, a capacity not to create *ex nihilo*, nor simply to repeat what has gone before, but to repeat *and* differ (*Difference and Repetition* being, of course, the title of one of the best known of Deleuze's later philosophical works). The beginning anew associated with desert islands in effect usurps the privilege of any single, unique and self-identical origin, appealing instead to a principle of originary difference that remains inassimilable to chronologically marked time: "In the ideal of beginning anew there is something that precedes the beginning itself, that takes it up to deepen and delay it in the passage of time. The desert island is the material of this something immemorial, this something most profound" (2004: 13–14).[2]

2 The notion of such a second beginning finds one of its most forceful expressions, for Deleuze, in the myth of the Flood, in which the Ark, as the crucible of a new beginning, finds its resting place on a *de facto* island, a solitary mountain peak protruding above the otherwise ubiquitous waters (Deleuze 2004: 13). In order to begin anew Deleuze suggests, it is necessary, at least provisionally, to separate oneself from whatever, according to the logic of chronological time-reckoning, came before. It is the water, the ocean or sea surrounding the island that embodies a principle of segregation necessary for such second creation to begin: "as though the island had pushed its desert outside" (2004: 11).

The Sea of the Dead

In the mythologies of land and sea invoked by Deleuze, the dead have often played a conspicuous role, both as named individuals and as an anonymous backdrop to the existence of the living. Elias Canetti, in his study *Crowds and Power*, suggests that the so-called "invisible crowd" of the dead may be amongst the most ancient and the most universal of human imaginings:

> Over the whole earth, wherever there are men, is found the conception of the *invisible dead*. It is tempting to call it humanity's oldest conception. There is certainly no horde, no tribe, no people which does not have abundant ideas about its dead. Man has been obsessed by them; they have been of enormous importance for him; the action of the dead upon the living has been an essential part of life itself. (Canetti 1992: 47)

Of these invisible dead, he writes: "generally it was assumed there were a great number of them" and he proceeds to give, by way of illustration, an eclectic range of ethnographic examples: the Bechuana of South Africa, who "believed all space to be full of the spirits of their ancestors"; the Chuckchi shamans of northeastern Siberia, with their "legions of auxiliary spirits" called upon in effecting cures; the belief, shared by the Sami of Northern Europe and the Tlingit of Alaska, that the *aurora borealis* or Northern Lights is composed of hosts of the dead; and the Gaelic term *sluagh*, used in the Scottish Highlands to refer to massed spirits of the dead who "fly about in great crowds like starlings" (Canetti 1992).[3]

Representations of these massed dead often depict them as a cloud or vapor, or else as a swirling liquid, overflowing all boundaries. For example, in the Afro-Cuban "inspiration" known as Palo or Palo Monte—a complex set of beliefs and practices relating to sorcery, healing, divination and communication with the dead—a key concept is the Ba-Kongo derived term *Kalunga*, the sea of the dead.[4] This is often used to invoke the dead as an undifferentiated and all-pervading substance, or as what Todd Ramón Ochoa, in a recent study, calls "the ambient dead." Ochoa writes that his own understanding of *Kalunga* (and his curiosity

3 The Scots Gaelic term *sluagh* ["spirit multitude"], Canetti adds, was often conjoined with another Gaelic term, *gairm*, meaning "to shout or cry." The resulting compound, *sluagh-ghairm*, meaning the battle-cry of the dead, is the origin of the Modern English "slogan" and Canetti notes (with apparent satisfaction) that "The expression we use for the battle-cries of our modern crowds derives from the Highland hosts of the dead" (Canetti 1992: 49).

4 Ochoa uses the term "inspiration" in part to avoid what he regards as the restrictively European implications of "religion" and in part to underscore Palo's orientation toward newness and creativity: "Inspiration implies a playful attitude toward past and future, as opposed to a perspective marked only by the trauma of dislocation and impossible recovery" (Ochoa 2007: 8).

about the role of the dead in Palo more generally) derives in large measure from conversations with Isidra, a 60-year-old woman and Palo practitioner living in an apartment building in the *El Cerro* district of Havana:

> Kalunga, as Isidra taught it through variegated reiteration, was the great, indifferent sea of the dead … As Isidra used it, Kalunga was not only haunted by the dead, but was also composed of the dead and simultaneously constitutive of them; the dead were immanent to it, in the same way a broth makes a soup. Kalunga, Isidra said, comprises all the dead that could possibly exist or have existed. It is ancient beyond memory, and within it the dead exceed plurality and become instead a dense and indistinguishable mass. According to Isidra, the world and experience, all things available to perception and perception itself, are a series of condensations within this fluid mass of the dead. (Ochoa 2007: 482)

In Isidra's rendering, Canetti's invisible crowd becomes something more akin to a universal primordial substance, a polymorphous stuff, capable of manifesting itself in a seemingly endless variety of forms (what Ochoa terms "versions" of the dead) and out of which not only material bodies but also thoughts and perceptions are fashioned, along with the "responsive dead"—the deceased loved ones, family members and, in some cases, historical figures called upon by Palo practitioners in the exercise of their craft. What *Kalunga* appears to name, then, Ochoa suggests, is a material power of self-differing out of which everything emerges yet that forever exceeds the actuality of what currently is. It is crucial, I think, that the dead referred to here "exceed plurality" becoming instead an indistinguishable and (by implication) uncountable mass. To affirm *Kalunga* as the immemorial sea of the dead is, one might say, to affirm reality as non-identical to and non-coincident with itself, to affirm a fundamentally unquantifiable power of becoming occupying not the status of a higher order of being or a transcendent creator god, but immanent rather to the very stuff of the universe. As such, the ever-present sea of the dead cannot be located within linear chronology or historical time, or, indeed, within any humanly conceived calibration of time. Instead, it affords an intimation of time as differentiating difference, a force at once creative and destructive, out of which humanly marked time and Palo's associated repertoire of stories and practices relating to the dead emerge and into which they recede. Modern Western philosophy has given a variety of its own names to this generative and dissipative time—Nietzsche's "eternal recurrence," Bergson's "duration," Deleuze's "pure and empty form of time" (Bergson 1991; Deleuze 1995; Nietzsche 2006). My concern here, however, is not to invoke these as explanatory or identifying labels but to consider some of the ways in which the widely recounted interplay between the particularized and the massed dead and their associated materialities can be understood as affording a distinctively revealing interface between humanly elaborated social-cultural worlds and the impersonal and inhuman powers of time and matter that are at once their condition of possibility and the source of their undoing.

The Dead of St Magnus

In terms of the distinction borrowed by Deleuze from geological science, the islands of Orkney can be assigned to the "continental" type. Their formation, however, long pre-dates the contours of present-day Europe. They are, rather, the remnants of a mountain chain formed between 400 and 600 million years ago by a collision between long-since dispersed prehistoric super-continents—the great southern continent of Gondwanaland and the great northern one of Laurasia. As the continents were forced apart by magma welling up from the earth's core—Europe separating from North America and the Atlantic Ocean opening up between them—the mountain ranges were eroded by sea to form the present-day island chains of Orkney and its northerly neighbor, Shetland—a process that continues to the present and that will eventually cause the islands to disappear beneath the waters from which they first emerged.

The earliest evidence of a human presence in Orkney dates from around 6500 BC. Successive waves of incomers followed, including the Neolithic builders of Knap of Howar, the oldest surviving house in Europe (occupied c. 3700–2800 BCE), on the northerly island of Papa Westray, of the Skara Brae settlement on Orkney Mainland (occupied c. 3100–2500 BCE) and of the stone burial cairns scattered across the Orcadian landscape. They in turn were followed from the eighth century by the Norse sea raiders, later turned farmers and fishermen, who became the dominant presence in Orkney, leading to the islands' annexation to the Kingdom of Norway in 875. The Earldom of Orkney, comprising Orkney and Shetland, was transferred to the Crown of Scotland in 1468 as part of a marriage settlement and became, in turn, part of the British nation state with the Acts of Union of 1707.[5] Today the islands of Orkney are home to some 19,000 people, most of them concentrated on Mainland, the largest island. The sea has continued to play an important role in the life of the islands, not only through commercial fishing, which declined sharply over the course of the twentieth century, but also through

5 Although the islands had come increasingly under Scottish influence during preceding centuries, not least through their predominantly Scottish clergy, the transfer of sovereignty was not intended, initially, to be a permanent one. Rather, it amounted to the pawning—or *Impignoration*—of the islands by Christian I of Norway and Denmark, who lacked the funds to provide an adequate dowry for his daughter on the occasion of her marriage to James III of Scotland. The ostensibly temporary nature of this arrangement has continued to be invoked by local activists. For example, in the 1980s campaigners against the disposal of radioactive material at the Dounreay power station in Caithness drew up a declaration calling on the Danish monarchy to challenge Orkney's constitutional status as part of the British Isles. Around the same time, the Orkney and Shetland Movement, a group seeking devolution for the northern isles, unsuccessfully contested the Orkney and Shetland Parliamentary seat in the 1987 general election, polling just over 3,000 votes. The islands' Nordic past has also been a persistent theme of Orcadian literature, including the work of Orkney's best known twentieth-century writer, the poet and novelist George Mackay Brown (1921–96), whose novel *Vinland (*1992) is set during the period of early Norse settlement (Hall 2010: 2–3; Thomson 2008: 189–205).

the onetime involvement of Orcadians in the whaling industry and, more recently, through the region's role in pioneering and developing marine energy technologies.[6] Over the centuries, however, the sea has been as much a cause of death as a source of livelihood, as winter storms and fast-running currents have claimed the lives of innumerable ships' crews. The wrecks submerged beneath Orkney's coastal waters were further added to by military vessels sunk during the First and Second World Wars, when Orkney was the headquarters of British Grand Fleet. Today these same wrecks are also a lucrative source of tourist revenues as parties of divers visit the islands every summer to explore them (Childe and Clarke 1983; Davidson 1989; Hacquebord 1990; Renfrew 1993; Thomson 2008; Towsey 2002; Watts 2012).

Let us turn our attention, however, from the sea to the *terra firma* of an island graveyard. The one in question adjoins St Magnus Cathedral in Kirkwall, the principal town of the Orkney Islands, situated on Orkney Mainland. Named after Orkney's patron saint and martyr, the building of the cathedral began in 1137 and was not completed for more than 300 years.[7] Its main entrance on Kirkwall's principal street is framed by two memorials, listing the names of Orcadians killed in the First and Second World Wars. Inside the cathedral, the remains of St Magnus himself, originally buried at nearby Birsay, are now interred inside a pillar. Surrounding him are a number of gravestones and memorials dating from the seventeenth century onward and featuring images of death along with inscriptions commemorating the deceased. More recent memorials include one to the poet and novelist George Mackay Brown (1921–96), arguably Orkney's most celebrated twentieth-century writer (Mooney 1947; Tarlow 1999).

St Magnus Cathedral and its grounds seem at first glance far removed from Isidra's vast sea of the dead. What we encounter here are not the anonymous hosts of the dead but rather those whose passing has been marked by funerary rites and whose presence on earth is materialized in the form of inscriptions, headstones, monuments and memorials—the transitoriness of fleshly substance replaced by an enduring body of stone. It would be a mistake, however, to draw too stark a

6 The European Marine Energy Center (EMEC) was established in Orkney in 2003 to provide testing facilities for developers of wave and tidal energy converters and currently operates 14 full-scale test berths, along with two "nursery" test sites for smaller devices or ones at an earlier stage of development. According to EMEC, Orkney was chosen as a base because of "its excellent oceanic wave regime, strong tidal currents, grid connection, sheltered harbour facilities and the renewable, maritime and environmental expertise that exists within the local community." In July 2012 the Pentland Firth and Orkney Waters were designated a Marine Energy Park (EMEC 2012).

7 Magnus was killed by his cousin Haakon Paulsson in or around 1115 in a dispute over succession to the earldom of Orkney. The story of his martyrdom is told in the Orkneyinga Saga, written in the early thirteenth century by an unknown Icelandic author (Palsson and Edwards 1981). Between 1154 and 1472, Orkney was part of the Norwegian archbishopric of Trondheim, after which it became part of the Scottish province of St Andrews. The cathedral was assigned to the people of Kirkwall by King James III of Scotland in a charter of 1486 and is owned today by the Royal Burgh of Kirkwall (St Magnus Cathedral 2012).

contrast between the massed and the individuated dead. Indeed, as Sarah Tarlow points out in her archaeological study of Orkney burial practices from the late sixteenth to the twentieth centuries, the degree and manner in which the dead were individuated varies considerably.[8] One obvious contrast is between burials in the churchyard and interments within the walls of the cathedral itself, the latter at one time the exclusive prerogative of the wealthy and powerful. There are significant variations too in the scale and design of the stones themselves (from the cross on a base flanked by a sword that is the earliest surviving monument, dating from the sixteenth century, to the mass-produced gravestones of the later nineteenth century), in the extent of the information given about the deceased (name, dates, details of occupation, relationship to surviving kin) and in whether the same stone commemorates a single individual or multiple family members (husband and wife, parents and children, etc.). The cathedral and churchyard attest also to the impact of the Protestant Reformation upon commemorative practices in Orkney. Protestant theology's repudiation of Purgatory and the possibility of intercession on behalf of the soul of the deceased led, Tarlow suggests, to a "radical dislocation" of mortuary and commemorative practices. Images of saints and crucifixes were henceforward proscribed (along with other representative religious art). At the same time, and in partial mitigation of the increased distancing of the living from the dead that the Reformation brought about, commemoration came to assume an increasingly didactic function, whereby the deceased could be presented both as a reminder to survivors of their own mortality and as an example of the virtuous Christian life that the living could emulate. The material record provides evidence of this shift in the seventeenth century and later bas relief memorials that are a conspicuous feature of the cathedral interior and that typically include a visual representation of the deceased, heraldic imagery emphasizing his or her social status, an inscription sometimes incorporating a didactic message and a generic image of death, often in the form of a skull or hourglass (Tarlow 1999: 50–107).[9] The one shown in Figure 1.1, dating from the eighteenth century, commemorates one Mary Young (d. 1756), wife of John Riddoch, a magistrate of Kirkwall ("She lived regarded and died regretted").[10]

8 Tarlow's study is based on records produced by the Orkney Graveyard Project, available for consultation in the form of a database in the Orkney Archive Office in Kirkwall (Tarlow 1999: 50–51).

9 The Reformed Church of Scotland also introduced a formal ban on intramural burials (on pain of excommunication) in 1579. Although this was strictly observed in many parts of Scotland, burials within the walls of St Magnus Cathedral appear to have continued until the early nineteenth century (Tarlow 1999: 89–94).

10 An adjacent memorial, commemorating George Liddell of Hamner (d. October 1681, aged 48) contains a more overtly didactic message (along with a family crest and skull): "Death hastens no flight avails to pay the tribute of mortality man is bound by nature's law all things go back to where they came and seek their mother and what has been returns again to nothingness But I lament the loss of time material damage everyone can make good no one can make good the loss of time."

Figure 1.1 Memorial to Mary Young, 1756, St Magnus Cathedral (photograph by author)

The image's power is attested by the fact that it continues to attract the attention of present-day visitors to the cathedral: the kneeling figure, her eyes raised heavenward, her hands clasped in prayer, framed by emblems of mortality, the origins of which extend back into the pre-Reformation Middle Ages and beyond. No doubt, as Tarlow's account so eloquently demonstrates, there is much to be learned both about and from it by reconstructing the initial circumstances of its production and reception. Yet perhaps there is also something here that escapes or exceeds the explanatory power of context. Doesn't the power of the image to derive in part from its lingering at the interface between the personalized deceased and the nameless and numberless hosts of the dead? Perhaps what continues to compel the spectator (the twenty-first century as much as the seventeenth-century one) is precisely the juxtaposition of a named individual with the still anticipated but inevitable death of everyone?

Their differences notwithstanding, it is tempting to see all such acts of commemoration as gestures of solidification aimed at pushing back the sea of the dead, just as the actual sea, which once washed close to the cathedral, has been pushed back over the centuries by a series of land reclamation projects (Thomson 2008). But is the sea of the dead really so far away? Decades or centuries of Orcadian weather—including persistent rain and winter storms—have rendered many of the inscriptions in St Magnus churchyard undecipherable. Others have been grown over by mosses and lichens, accretions of animate matter transforming incised text into a semi-abstract surface of green and grey, out of which, occasionally, there emerges into legibility a date, name or term of relatedness—husband, wife, daughter, son. Those able to afford the distinction of an intramural burial have often fared little better.[11] Renovations to the cathedral during the late nineteenth century and early twentieth centuries resulted in the wholesale removal of many of the remains interred beneath the floor of the nave and their reburial *en masse* in a pit outside.[12] At the same time, all those who lie in the cathedral and churchyard are underlain by a host of earlier occupants whose presence is no longer recorded, while the Christian burials housed in the vicinity of St Magnus are surrounded by traces of the now anonymous, pre-Christian dead in the guise of the Neolithic chambered tombs that dot the Orcadian landscape, like the one known as Maes Howe, situated around 15 miles west of Kirkwall (Ashmore 1988).

11 The practice of interments inside the cathedral appears to have continued until the early decades of the nineteenth century, causing the level of the floor in the nave to be continuously raised. One description from 1847 refers to the "effluvia" unleashed by the opening of these graves to facilitate restoration works (Tarlow 1999: 93).

12 Albert Thomson, who served as Curator of St Magnus Cathedral for 32 years, recalls a conversation with an eyewitness to the reburial of the intramural dead, an elderly Kirkwall man ("with several drinks taken") who looked into the hole before it was filled in and remarked: "Good God, boys, there be some scramble here at the Resurrection!" (OLSA 386).

Where Do They Come From?

It would seem then that, from the perspective of the *longue durée*, gravestones and memorials are no more than temporary expedients and that it is the fate of the (more or less) individuated dead to be absorbed back into the undifferentiated mass of their forebears, like the Mongolian shamans described by Caroline Humphrey, who are believed after death to be amalgamated, gradually, with the anonymous powers animating the landscape in which they once lived (Humphrey 1995). Where do they come from then, these hosts of the dead—swarming in the depths of the sea, billowing in the air, shimmering in the boreal night sky? Are they simply humanity's accumulated past, the aggregate of all those who have lived and died and slipped from the memory of those still living? Many narratives of encounters between the living and the hosts of the dead seem to suggest just such a possibility. Think, for example, of the Europe-wide repertoire of stories relating to the Wild Hunt—a raucous, nocturnal, sometimes airborne animal–human cavalcade in which spectators have sometimes claimed to identify the figures of recently deceased relatives, new recruits as it were to the massed ranks of the dead, who have not yet sunk wholly into the anonymity of forgetting (Davidson 2001). Do we not find here an explanation of how the invisible crowd of the dead has been constituted and added to over the course of human history by the sum total of individual deaths?

Marcel Granet, writing of the religion of ancient China, suggests a different possibility. Granet argues that the primacy of the collective over the individual—a view that underpinned the sociology of his onetime teacher, Émile Durkheim—applied no less to the relationship between the dead and the living. The vision of the dead as an amorphous and undifferentiated mass was, in other words, prior to any conception of the named and individuated dead. The dead, however, in Granet's account, feature less as a collective representation than as a primordial substance. Granet finds the earliest religious conceptions to be embodied in the ancient beliefs and practices of the Chinese peasantry. A central place in peasant religion was occupied by the Earth as a source of fecundity. A bond of consubstantiality linked the Earth to its human occupants and a similar bond conjoined the living and the dead. The dead were subject to a double ceremony of burial. Initially they were interred within the family compound for the time it took the flesh to rot, so that "the stuff of the dead entered the domestic Soil" after which their remains were transferred to an ancestral graveyard forbidden to all who were not family members. The view thus arose, Granet suggests, that the conception of new life was the work of fecund powers emanating from the soil where the dead themselves became absorbed into the maternal substance of the Earth. The living were understood to be linked to the dead through their shared participation in this amorphous ancestral substance. Unlike the living, however, the dead were not differentiated: "The family was divided into two parts: one was that of the living, strongly united but endowed with the peculiarities inherent in each individual life; the other, that of the dead, formed an indistinct mass." Only at a later time, with the development of feudalism, was the individuated character of the living transferred to the dead as particularized ancestors to whom offerings could be made and acts of worship addressed. Such a conception of specific and identifiable ancestors was, however, distilled out of an older sense of the dead as an undifferentiated ambient presence, "floating confusedly in a corner of the house" (Granet 1975: 52). If we accept Granet's claim regarding the primacy of the ambient dead, what requires explanation is not the ontogeny of Canetti's invisible crowd but rather the way in which the individualized dead—and, perhaps, the living themselves—are condensed out of it. In other words, are ancestral cults, death ritual and practices of memorialization understandable as the means by which a distinguishable presence is solidified out of and thus disembedded from the always antecedent throng of the unnamed and undifferentiated dead?

Anthropological literature, at least since Hertz, has frequently emphasized the role of death ritual in effecting a definitive separation between living and dead, consigning the latter to their own realm (what Hertz calls a "society of the dead") where relations with them can be undertaken without risk to the living (Hertz 1960: 58). It is worth asking, however, whether such an undertaking has not often required, at the same time, a certain self-conscious differentiation of the dead that is as much a material practice as an act of symbolization. In order for relations with them to be regulated and managed, the dead must be produced as something other than an all-pervading and formless substance. Certainly the widely documented practice of exhuming and reburying the remains of the dead once the flesh and other decomposable parts of the body have rotted away can be understood as allowing the liquid indeterminacy of putrefaction to be replaced by the solidity of dry bones, which can then be safely consigned to their final resting place. The multifarious material culture of death practices—including tombs, ossuaries, memorials and grave goods—needs to be understood too as concerned not only with symbolizing the deceased's social status or of the community's beliefs about the afterlife but also with intervening in and modifying states of matter. Hertz's account intimates as much in its extended discussion of the threat of pollution widely ascribed to the remains of the dead during the intermediary period between their initial and final disposal. Hertz writes of the "impure cloud" that surrounds the deceased, polluting everything it touches, including not only people and objects that have been in physical contact with the corpse but also everything connected with the deceased in the minds of survivors (1960: 38).[13] If Hertz ultimately ascribes these perceived dangers to the threat to the social collective posed by the death of one of its members (particularly an individual such as a chief in whom the power and endurance of the collective is understood to be embodied), his formulation here evokes a pervading and miasmic materiality more akin to Canetti's invisible crowd or Palo's sea of the dead than to his own society of the dead, modeled as the latter is on the institutions and practices of the society of the living, albeit freed from the constraints of mortality. The taboos and gestures of ritual prophylaxis described by Hertz here seem directed at a presence

13 Hertz's examples here (and throughout his study) are drawn predominantly from the Olo Ngaju, a grouping of riverrine peoples in southeastern Borneo (Hertz 1960: 29).

more palpable—if no less elusive—than the "social stuff" that Hertz's mentor, Durkheim, has recently been criticized for assuming as the defining constituent of human collectives (e.g. Latour 2005). Instead, Hertz's account points on occasion to the powers and dangers of a very different sort of "stuff"—an other than human materiality that is, nonetheless, profoundly implicated in the fashioning of bodies, selves and collective practices even as it overruns and exceeds their bounds and determinations. Perhaps then what is at stake in death practices is not so much the objectification of social relations or values as objectification in and of itself, for its own sake—a process that is at once framed within and staked against the amorphous liquid or vaporous mass of the dead that, as Canetti pointed out, has so ubiquitously haunted the human imagination.

The Time of the Ancestors

Figure 1.2 Cuween Hill Neolithic tomb, near Finstown, Orkney Mainland (photograph by author)

The Neolithic settlers who were the builders of Orkney's chambered burial cairns arrived in the islands around 5,500 years before the present, probably crossing Pentland Firth in skin boats from the vicinity of Caithness on the Scottish mainland (Ritchie 1984). In addition to farmsteads and dwellings, they also built structures to house their dead. These vary in the details of their construction, but characteristically take the form of a stone cairn enclosing a rectangular chamber, sometimes subdivided by upright slabs and accessed by a low, narrow passage requiring the visitor to stoop or, in some cases, to crawl. Archaeologists have pointed out parallels between the layout of the chambered burial cairns and that of surviving Neolithic dwellings like Knap of Howar on the island of Papa Westray or the houses comprising the village of Skara Brae on Orkney Mainland, suggesting that the tombs were conceived of by their builders as "houses" for the dead. This in turn has fueled speculation that the Neolithic in northwest Europe (and elsewhere) marks a shift in attitudes toward the dead and, by extension, the emergence of a new conception of time, both of which are materialized in the archaeological record. Alasdair Whittle, for example, writes that the Neolithic period ushered in what he calls "the time of the ancestors"—a conception combining a sense of continuity between past and present, living and dead, with a new awareness of lineal descent whereby the living were able to articulate their own ancestry and their own constitution as a descent group via a sequence of forebears culminating in the present generation. It was, he suggests, the need to accommodate the remains of these newly differentiated collective ancestors that prompted the building of tombs as houses for the dead, including the chambered cairns that remain such a visible feature of Orkney's present-day landscape (Whittle 1996: 1, 355, 365–6).

Nonetheless, if recent archeological scholarship has sometimes been inclined to find in Neolithic mortuary architecture the expression of a newly differentiated view of the ancestral dead, coupled with an emergent sense of lineal chronology, its findings have also demonstrated the extent to which the efforts of Orkney's Neolithic settlers to objectify and commemorate their dead have been effaced or undone by the millennia that have passed between the building of the tombs and latter-day archaeological interpretations of them. For all the sophistication of current techniques of dating and analysis, the dead whose remains were housed in Orkney's chambered cairns remain resolutely anonymous. Indeed, the material record presents an often confusing and uncertain picture with regard even to the deposition of remains. The largest and best known of Orkney's burial cairns, Maes Howe, situated between Kirkwall and Stromness, has not been found to contain any human remains beyond a few skull fragments intermixed with animal bones. In contrast, Mid Howe, on the west coast of the island of Rousay, housed the remains of no fewer than 25 individuals, some arranged in crouched burials, others grouped together in a heap of bones at the center of the main chamber. Yet other tombs have been found to contain substantial quantities of animal as well as human remains, like Cuween Hill, lying southeast of the present-day settlement of Finstown on Orkney Mainland, where the remains of eight human skeletons were found along with 24 dog skulls, prompting speculation that the latter represent a sacrificial offering or the expression of a totemic complex (Davidson 1989). It has therefore proven impossible to arrive at general conclusions as to whether, for example, Orkney's chambered cairns were indeed the final resting place of the dead, whether they afforded a temporary lodging during the process of decay and decomposition or whether remains were moved in and out of them at intervals in the course of ritual activity. In the case of Orkney's prehistoric inhabitants, one effect of this hindsight-induced uncertainty has been to blur any clear-cut distinction between the collectively recognized ancestral dead and the always prior and undifferentiated mass of the dead.

If archaeological interpretations of sites like Maes Howe have often tended to see them as marking a move away from notions of an undifferentiated sea of the dead, it is nonetheless arguable that the actual sea that washes against Orkney's shores remained a powerful presence in the cosmologies of the region's Neolithic inhabitants. Chris Fowler and Vicki Cummings have suggested that the interplay of water and stone was a central theme in Neolithic cosmological thinking. Although the focus of their discussion is the eastern side of the Irish Sea (south- and northwest Wales, the Isle of Man and southwest Scotland) they argue that a close association between the two substances is to be found throughout the Neolithic world, finding one of its principal manifestations in chambered tombs, understood as places of transformation and transition, both between the realms of the living and the dead and between land and sea. They point to the location of many such tombs adjacent to the shoreline or on headlands or promontories, to the fact that stone from intertidal zones was often used in their construction, to the frequency of deposits of shells and pebbles and to the decorative and depositional use of quartz as a material commonly found both in inland, mountainous regions and in close proximity to the sea. For Fowler and Cummings, drawing on analogies with recent ethnographies of Melanesia, the conjunction of water and stone that characterizes the construction and location of many Neolithic chambered tombs is revealing of a conception of personhood and of the human life-course as unfolding via a series of transformations, finding one of their privileged expressions in the transition between solid and liquid modalities of matter and culminating in death, where the sea represents perhaps the final destination of the deceased (Fowler and Cummings 2003: 16).

Although some of Orkney's chambered cairns, including Maes Howe, are built at a distance from the surrounding sea, others stand in close proximity to it, including Isibister on South Ronaldsay and Mid Howe, Knowe of Yarso, Blackhammer and Taversoe Tuick on the south coast of Rousay, while Unstan, near Stromness on the south coast of Mainland, is built on a promontory extending into the Loch of Stenness. References to the sea are also to be found among the grave deposits in many of Orkney's tombs, which include the remains of whales and other marine animals and, in the case of Isbister, the bones and talons of sea eagles, giving the site its alternative name, 'The Tomb of the Eagles' (Hedges 1984).

Even when not explicitly referenced in the archaeological record, the sea is never far away: in fact, it's getting closer all the time! The Neolithic village of Skara Brae, occupied between 3180 BC and around 2500 BC, is the best surviving and probably the most extensively researched prehistoric settlement in northern Europe.[14]

14 Since its rediscovery, Skara Brae has claimed the attention of a succession of internationally prominent archaeologists, including V. Gordon Childe, Stuart Piggott and Colin Renfrew (Childe 1931; Piggott 1954; Renfrew 1984).

Figure 1.3 Skara Brae, Orkney Mainland (photograph by author)

Originally it stood at some distance from the sea, separated from it by sand dunes, which eventually engulfed the settlement after it was abandoned by its inhabitants for reasons that remain obscure, only to be uncovered in the mid nineteenth century when much of the sand was dispersed by a violent storm. Today, the remains of the village, comprising a number of houses and other structures interconnected by tunnels, sit precariously at the water's edge. In response, the conservators of the site, Historic Scotland, have closed off access to the dwellings themselves, in order that the effects of ongoing marine erosion should not be compounded by the feet of visitors. Nonetheless, given the continuing and unstoppable encroachment of the sea, the future of Skara Brae remains at best uncertain (Childe and Clarke 1983).

"Our Island Home is Doomed"

Perhaps unsurprisingly, the encroachment of the sea upon the land has been a persistent theme of Orcadian imaginaries of place. Stromness, Orkney Mainland: Walter Traill Dennison (1825–94), probably Orkney's best known and most prolific nineteenth-century folklorist, is addressing a meeting of the Orkney Natural History Society. Dennison was the son of a well-to-do farming family from Sanday, the largest of Orkney's northern isles, lying around 20 miles northeast of Orkney Mainland. His lecture, published in 1893 in the journal *The Scottish Antiquary* is devoted to the effects of marine erosion on Orkney's landscape and archaeological heritage, including graves and memorials to the dead, and draws repeatedly on examples from his native island. He begins, however, not with burials of deceased humans but with the remnants of submerged prehistoric forests. On the west side of the bay of Otterswick, in Sanday, an area now accessible only at the low water line during the spring tide in March, it is possible, he points out, to dig through the overlying layer of sand to reach a bed of moss "in which the decaying skeletons of trees lie in every conceivable position" (Dennison 1995: 171). Such submerged forests, Dennison notes, can be found too on many of the other islands: at Storehouse Bay on the island of North Ronaldsay; at Pierowall on Westray; at Millbay on Stronsay; and on Orkney Mainland and the south isles, affording proof that the land on which the trees once grew was formerly above sea level and that the islands of the Orkney archipelago have been undergoing a gradual process of subsistence (1995: 172). Dennison offers a brief imaginative reconstruction of these now vanished landscapes—the tree trunks and spreading leaves, the animals that once roamed there and the birds that sang in the branches—before going on to enumerate the traces of human habitation that are in the process of being effaced by the encroachments of the sea—as attested by the fragmentary ruins of houses and kelp stores at Swarthammer and Hamness, the side walls and seaward gables of which have now disappeared beneath the waters (1995: 171–2, 185).

Halfway through the lecture though, the perspective abruptly shifts. Dennison asks his audience to reflect on the transitoriness of all terrestrial forms, considered from the vantage point of geological time:

> In our "talk, talk", we speak of the firm, stable, immovable earth. No words can be more fallacious when applied to our world. The truth is, we live on a globe of fire, ready at any moment, from many causes, to be hurled into terrific destruction. Imagine a huge chasm opening longitudinally in that part of the abysmal sea on which the two Atlantic oceans rest or roll. Those oceans would at once be precipitated on a tremendous mass of igneous matter, whereby an amount of steam would be generated sufficient to explode the world, making the earth like a bursting bombshell … And while we talk of the immutable laws of nature, we should remember that all these laws must succumb to the inexorable law of exhaustion. (1995: 173)

From an account of transformations of the familiar landscape, we are catapulted into a vision of planetary destruction evoking the obliteration not only of the human presence on earth but also of the terrestrial globe itself as the ecological support of all life.

It is striking that Dennison should be propelled toward such a vision precisely by consideration of the interplay—the "strife"—between land and sea in an island setting. Islands appear here as sites not only of new beginnings, or beginnings anew—as they were for Deleuze—but also of violent and catastrophic endings. Concrete instances of death—whether it be the demise of prehistoric forests or of the human inhabitants of the ruined houses scattered along Sanday's shoreline—

are absorbed first of all into the encroaching waters of the North Sea and, finally, into a speculative conjuration of cosmic flux and dissolution capable of eradicating all vestiges of a human presence. In his subsequent writings, including his collaborations with Guattari, Deleuze would re-articulate the elemental strife of land and sea in terms of what he called a "geophilosophy," situated in relation to the earth as the inhuman ground of human thought and action and tracing a continuous two-way traffic between the territorializing inscription of the earth by history, language and politics and the de-territorializing impetus of the earth itself as the undoing and the virtual limit of any possible actualization of territory and thus of any humanly conceivable project of world-making (Deleuze and Guattari 1994: 85–113; Flaxman 2012: 72–114). More recently, the philosophical imperative to envision a cosmic reality from which humanity is radically absent has been re-asserted by Quentin Meillasoux, who refers to such a reality as the realm of the "ancestral" (or the "Great Outdoors"), referring not to the human ancestors documented in archaeological and folklore scholarship but to that which pre- or post-dates humanity and that demands as such to be thought as existing independently of its being given to a knowing, perceiving human subject. For Meillasoux it is, pre-eminently, modern science, including geology, that has provided access to ancestral realities such as the formation of the earth and the beginnings of terrestrial life, in the process throwing down a challenge to post-Kantian philosophy's insistence on the inescapably contextual, linguistic or subjectival mediation of all knowledge (Meillasoux 2008: 1–27). Yet to suggest, as Meillasoux appears to, that knowledge of the ancestral is the exclusive prerogative of a self-sufficient philosophical or scientific rationality risks obscuring the ways in which the antecedent inhumanity of the ancestral remains implicated in human thought and language, no less than in the perhaps more self-evident materiality of bodies and perceptions. If the disavowal of any possibility of direct knowledge of reality in-itself has been a defining gesture of modern, self-styled "critical" philosophies, it is surely worth asking whether intimations of a reality existing independently of our capacity as humans to think or perceive it have not previously found expression through other genres and media—whether, indeed, such expressions might not be as old as humanity, as old as "culture." Might not what has been referred to as "culture" be nothing more or less than the human appropriation and encoding of an inhuman, anorganic force of matter that forever exceeds and displaces the humanly constructed orders fashioned from it?[15] If so, might it not be the case that this displacement from within and without has attained one of its most potent manifestations in the immemorial and unresolved strife of land and sea and its associated imaginaries?

The ever present and ever encroaching sea that advances upon Orkney's islands was an abiding preoccupation of Dennison's output as a folklorist.

15 Clare Colebrook has recently argued that Deleuze conceives of "Life" in precisely these terms and that he is therefore not to be taken as a "vitalist" in the sense understood by some of his recent critics (Colebrook 2010).

Dennison's writings on the Orcadian folklore of the sea first appeared as a series of articles in *The Scottish Antiquary* between 1890 and 1893. Here the sea appears not only as a constant presence in the life of the islands but also as the abode of a variety of other-than-human powers and agencies. These included: the benign and maternal Sea Mother ("Mither of the Sea") who, each spring, fought and subdued her antagonist Teran (the bringer of winter storms) to ensure calm weather during the summer months; the Stoor Worm, a variety of sea serpent; the Nuckelavee, or Devil of the Sea, whose venomous breath blighted crops and caused animals to sicken; the misshapen Sea Trow; the Fin Folk, a race of beings usually described as human in outward appearance but living beneath the waves in and occasionally abducting human spouses to join them; and, perhaps best known, the selkies, or seal people, beings residing in the sea in the guise of seals but sometimes venturing ashore at twilight to cast off their seal skins and dance naked in human form on the sand.[16] For Dennison, the repertoire of stories about these and other denizens of the sea that he collected from his native island of Sanday and elsewhere represented a series of cultural strata (or "survivals" in Edward Tylor's influential phrase) extending back to the earliest human conceptions. The sea was thus not only the resting place of countless dead of past generations but also a repository of archaic imaginings relating to other than human forces and presences. Like Tylor, Dennison suggested that in these vestigial traces could be read a history of the evolution of human thought and imagination. Beginning with the Sea Mother ("answering to all the purposes of protoplasm") he traced a progression through Teran and the Stoor Worm via Nuckelavee ("half man and half beast"), the Sea Trow ("in the form of a man, with the mind of a beast"), the seal ("a beast, yet able to assume the form of a man") to the Fin Folk ("with astute mind, and well developed human form, yet with all the conveniences of a fish for aqueous existence and locomotion"), a progression tending, he suggested, toward the increasingly anthropomorphizing personification of elemental forces (Dennison 1995: 30). Dennison's descriptions suggest, however, the possibility of considering the folklore of the sea as something more than the projection of human-identified traits onto an inanimate geography. The contemporaneous co-existence (at least until the late nineteenth century) of the various storied beings inventoried by Dennison could be understood as indexing not the linear temporality underpinning Victorian stadial theories of social evolution but rather a time more akin to that of Deleuze's pre- and post-historical strife of land and sea, of the self-differentiating potentiality of time and matter to which folkloric imaginings of elemental powers continuously refer us. The Orcadian folklore of the sea, to which Dennision devoted a lifetime of scholarly labors, might be approached then, not as the evidential scaffolding on which to hang a unidirectional account of human progress, but as an exemplifcation of what the young Deleuze identified as the "mythic" imagination, a collectively exercised power of fabulation that

16 On Orcadian folklore of the sea, see also Marwick (2000).

appropriates, condenses and elaborates the other-than-human forces and materials out of which human worlds are at once formed and carried beyond themselves.[17]

"The Sea is the Land's Edge Also"

It is worth, I think, setting the reiterative presencing of the sea in the guise of folklore, material culture and marine erosion against, on the one hand, the sometimes voiced archeological interpretation of Neolithic tombs as marking the differentiation of the dead as a linearly unfolding sequence of group ancestors and on the other, the more varied technologies of particularization (gravestones, memorials, etc.) in evidence in the churchyard of St Magnus Cathedral. As the unevenness and variability of the intervening archaeological record reminds us, we should be wary of locating such divergent expressions in a single trajectory of progressive individuation extending from the prehistoric past to the early twenty-first century present. What is evoked rather by Dennison's anticipated cataclysm, by the fate of sites such as Skara Brae and by the material record itself is not a linear narrative of historical progress but a persistent and unresolved two-way movement between differentiation and de-differentiation, between, on the one hand, the solidification of collective, individual or sometimes named presences through the construction of houses for the dead, the ritual processing of remains, the erection of gravestones and monuments, the carving of inscriptions and, on the other hand, the continuing and finally unstoppable dissolution of the same back into the amorphous, all-pervasive and always prior backdrop so powerfully evoked in Isidra's exposition of the Palo term *Kalunga* as "the great indifferent sea of the dead" (Ochoa 2007: 482). Further, are not the sea of the dead, Canetti's invisible crowd and the more familiar, terrestrial sea from which the former borrows its name, themselves condensations out of the immense, indifferent and always antecedent matter of the universe—of planets and stars, solar systems and galaxies that were born and died long before the appearance of life on earth? If the cultural death-work performed by funerary rites, inhumation practices and acts of memorialization is apt to appear always as an attempt to resist or slow the dissolution of the deceased into the generative and un-totalizable mass of what came before (and will come after), these death practices nonetheless affirm and make visible at the same time the inevitability of such dissolution. What such gestures produce then is, arguably, a transitional space between differentiation and assimilation, one in which the striving toward commemoration or perpetuation becomes, at the same time, the disclosure of ephemerality, of our participation as humans and culture-makers in a materiality that at once constitutes and dispossesses us. Our bodies will liquefy, our stone monuments will crumble, our inscriptions will fade. Are not we who call

17 For an anthropological account of creativity that parallels Deleuze's, see McLean (2009). For an extended discussion of the notion of "fabulation" as it features in Deleuze's writings, see Bogue (2010).

ourselves the living, along with our tombs, memorials, funerary rites and prayers for the dead, no more than a passing hesitation, a transitory holding back in the face of the summons that murmurs continuously and irresistibly from the vast sea of the dead? Perhaps, indeed, it is the privileged role of the dead, the anonymous and mythic dead as much as our deceased loved ones, friends and family members, to remind us continuously of this. Dennison concludes his address to the Orkney Natural History Society by returning to a human time-scale, albeit one poised precariously on the brink of its own dissolution. He reminds his audience that their own memorials to the dead, the solid ground on which they stand and, indeed, they themselves, are, likewise, no more than a fleeting presence, fated to be reclaimed by the sea that beats continuously against Orkney's shores:

> our island home is doomed. In a few short ages the lobster and crab will crawl on our cold hearthstones; whales and fishes will disport above where our chimney tops now reach; seaweeds and limpets will grow on our gravestones, and our graves will be nowhere. But our dust will be safe in that most glorious of all sepulchers – the mighty ocean – on which "time writes no wrinkle." (Dennison 1995: 186)

References

Ashmore, Patrick. 1988. *Maes Howe*. Edinburgh: Historic Buildings and Monuments, HMSO.

Bergson, Henri. 1991 [1908]. *Matter and Memory*. New York: Zone Books.

Bogue, Ronald. 2010. *Deleuzian Fabulation and the Scars of History*. Edinburgh: Edinburgh University Press.

Canetti, Elias. 1992 [1960]. *Crowds and Power*. London: Penguin Books.

Childe, V. Gordon. 1931. *Skara Brae: A Pictish Village in Orkney*. London: Kegan Paul, Trench, Trubner & Co.

Childe, V. Gordon and D.V. Clarke. 1983. *Skara Brae*. Edinburgh: HMSO.

Colebrook, Clare. 2010. *Deleuze and the Meaning of Life*. London: Continuum.

Davidson, Hilda Ellis. 2001. The Wild Hunt, in *Supernatural Enemies*, edited by Hilda Ellis Davidson and Anna Chaudhri. Durham, NC: University of North Carolina Press, 163–76.

Davidson, James L. 1989. *The Chambered Cairns of Orkney: An Inventory of the Structures and Their Contents*. Edinburgh: Edinburgh University Press.

Deleuze, Gilles. 1995 [1968]. *Difference and Repetition*. New York: Columbia University Press.

Deleuze, Gilles. 2004. *Desert Islands and Other Texts, 1953–1974*. Cambridge, MA: MIT Press.

Deleuze, Gilles and Félix Guattari. 1994. *What is Philosophy?* London: Verso.

Dening, Greg. 1980. *Islands and Beaches: Discourse on a Silent Land: Marquesas, 1774–1880*. Honolulu: University of Hawaii Press.

Dennison, Walter Traill. 1995. *Orkney Folklore and Sea Legends: Studies of Traditional Life and Folklore*. Kirkwall: The Orkney Press.

EMEC. 2012. European Marine Energy Council. Available at: www.emec.org.uk [accessed 08/01/2012].

Finlayson, Bill and Kevin Edwards. 2003. The Mesolithic, in *Scotland After the Ice Age: Environment, Archaeology and History, 8000 BC – AD 1000*, edited by Kevin J. Edwards and Ian B.M. Ralston. Edinburgh: Edinburgh University Press, 122–3.

Flaxman, Gregory. 2012. *Gilles Deleuze and the Fabulation of Philosophy*. Minneapolis: University of Minnesota Press.

Fowler, Chris and Vicky Cummings. 2003. Places of Transformation: Building Monuments from Water and Stone in the Neolithic of the Irish Sea. *Journal of the Royal Anthropological Institute*, 9(1), 1–20.

Granet, Marcel. 1975 [1922]. *The Religion of the Chinese People*. New York: Harper & Row.

Hacquebord, Louwrens. 1990. "There She Blows": A Brief History of Whaling, in *Whaling Communities*, edited by Elizabeth Verstergaard. *North Atlantic Studies*, 2(1–2), 11–20.

Hall, Simon J. 2010. *The History of Orkney Literature*. Edinburgh: Birlinn.

Hedges, John W. 1984. *Tomb of the Eagles: A Window on Stone Age Tribal Britain*. London: John Murray.

Hertz, Robert. 1960 [1907]. *Death and the Right Hand*. Glencoe, IL: The Free Press.

Humphrey, Caroline. 1995. Chiefly and Shamanist Landscapes in Mongolia, in *The Anthropology of Landscape: Perspectives on Space and Place*, edited by Eric Hirsch and Michael O'Hanlon. Oxford: Clarendon Press, 135–62.

Latour, Bruno. 2005. *Reassembling the Social: An Introduction to Actor-Network Theory*. New York: Oxford University Press.

Malinowski, Bronislaw. 1984 [1922]. *Argonauts of the Western Pacific: An Account of Native Enterprise and Adventure in the Archipelagoes of Melanesian New Guinea*. Prospect Heights: Waveland Press.

Marwick, Ernest. 2000. *The Folklore of Orkney and Shetland*. Edinburgh: Birlinn.

Meillasoux, Quentin. 2008. *After Finitude: An Essay on the Necessity of Contingency*. London: Continuum.

McLean, Stuart. 2009. Stories and Cosmogonies: Imagining Creativity beyond "Nature" and "Culture." *Cultural Anthropology*, 24(2), 213–35.

Mooney, James. 1947. *The Cathedral and Royal Burgh of Kirkwall*. Kirkwall: W.R. Mackintosh.

Nietzsche, Friedrich. 2006 [1883–5]. *Thus Spoke Zarathustra*. Cambridge: Cambridge University Press.

Ochoa, Todd Ramón. 2007. Versions of the Dead: Kalunga, Cuban-Kongo Materiality and Ethnography. *Cultural Anthropology*, 22(4), 473–500.

OLSA. Orkney Library Sound Archive. Numbers refer to tapes of recorded interviews.

Palsson, Herman and Paul Edwards. 1981. *Orkneyinga Saga: The History of the Earls of Orkney*. London: Penguin.

Piggott, Stuart. 1954. *Neolithic Cultures of the British Isles*. Cambridge: Cambridge University Press.

Sahlins, Marshall. 1985. *Islands of History*. Chicago: University of Chicago Press.

Renfrew, Colin (ed.) 1993. *The Prehistory of Orkney*. Edinburgh: Edinburgh University Press.

Ritchie, Anna. 1984. First Settlers, in *The Prehistory of Orkney*, edited by Colin Renfrew. Edinburgh: Edinburgh University Press, 36–53.

St Magnus Cathedral. 2012. Available at: http://www.stmagnus.org/history.html [accessed 14/3/2012].

Tarlow, Sarah. 1999. *Bereavement and Commemoration: An Archaeology of Mortality*. Oxford: Blackwell Publishers.

Thomson, William P.L. 2008. *The New History of Orkney*. Edinburgh: Birlinn.

Towrie, Sigurd. 2008. New Contender for Orkney's Oldest Settlement Site. *Orkneyjar*, 20 March. Available at: http://www.orkneyjar.com/archaeology/stronsayflints2008.htm [accessed 15/1/2012].

Towsey, Kate (ed.) 2002. *Orkney and the Sea: An Oral History*. Kirkwall: Orkney Heritage.

Watts, Laura. 2012. OrkneyLab: An Archipelago Experiment in Futures, in *Imagining Landscapes: Past, Present and Future*, edited by Monica Janowski and Tim Ingold. London: Ashgate, 59–76.

Wickham-Jones, Caroline. 1994. *Scotland's First Settlers*. London: B.T. Batsford/Historic Scotland.

Wickham-Jones, Caroline. 2007. *Orkney: A Historical Guide*. Edinburgh: Birlinn.

Whittle, Alasdair. 1996. *Europe in the Neolithic: The Creation of New Worlds*. Cambridge: Cambridge University Press.

Chapter 2
Death, Witchcraft and the Temporal Aspects of Divination

Knut Rio

Introduction

It is curious that even though the technologies of death—i.e. divination, witchcraft and sorcery—expressively work through materially mediated practices, their objects often have no particular character or aesthetic value. The case is that divination, the work of discovering the cause and agency behind death or illness, achieves this through manipulation of objects, but the objects are often of extraordinary simplicity or minimalism. They can be stones or shells picked up from the ground and tossed in the air, marks made in the sand, leaves put into hot water, water put in a bowl, or plain pieces of wood held in particular ways. Likewise, remedies for sorcery can be just remains of someone's hair, fingernails or pieces of clothing. Artifacts of witchcraft can be ashes of human bone, a needle or a particular thorn or some other hardly observable thing. Divination, sorcery and witchcraft seem to conceal their practices and potency in this everyday, simplistic character of their work and instruments. Both the diviner and the sorcerer work on the basis of a technology that is *next to nothing* even though they bear great results in terms of life and death. But how come pebbles and thorns are chosen for creating miracles?

I shall argue that this has to do with the very character of these practices as operating with an intentionality toward a realm that is also next to nothing, if not nothing at all. In order to approach this realm, such as approaching the witch who is constituted in this realm and bursting forth from it, the diviner has to handle materially the very bits and pieces of the most ordinary of the everyday things; items that are closer to nothing than anything else. With regard to the *temporality* of death, I believe this goes in parallel with its materiality. The diviner picks up some stones, shell or sticks from the ground in order to let these elementary forms reorient themselves and illustrate the composition of elements and personnel in a different realm, and time is restructured in the same sense. In order to recall a moment of death, the diviner also has to toss time up in the air, and to let it restructure itself so that it can become comprehensible for those who need to know about the realm of death itself.

This chapter thus offers preliminary thoughts on how divination works to recall bygone moments in time, related to the potent moment when someone has died or become ill from witchcraft or sorcery. I trace the causation and logic of witchcraft trials and look at how one is seeking evidence by so to say walking backwards into a past that was never a present. In terms of material mediation and time, the theme of this volume, I look at how beating or killing a witch, like seizing into the inner organs of a dead animal in divination practices, is creating an intermediary space for communication with spirits, objects and persons that all help to document a different realm of reality. In this frame of understanding divination seeks into the space of death by manipulating elementary material objects, so that they are made to tell the truth about the space beyond life and beyond materiality, beyond the realm of being-here and being-now.

This means I run a general claim about witchcraft, however awkward that might be, based upon my experiences in the Melanesian nation of Vanuatu. I agree that phenomena of the occult must be understood in relation to their rootedness in cultural-historical and structural circumstances, but I also see the need to articulate universal connections in the way people around the world act upon death and illness, and how they with intentionality construct and act upon that realm of life that is not life. This chapter is about the ontological ambivalence and ambiguity that witchcraft reveals, and it tries to get closer to its social status as an event. Even though people around the world build whole institutions around witchcraft as an *existence*, there is often simultaneously the idea that it is "not here, not now and not on my part." It is always *not there* in the moment, but always threatening to *become* as a potentiality, and, further, to *having been* in after-the-fact demonstrations by diviners or oracles. This is maybe why the word for sorcery among the Sinhala Buddhists of Sri Lanka is derived from the Sanskrit *zunya*, meaning "nonexistence, vacuum, nonentity" or "the void from which existence springs" (see Kapferer 1997: 37). My explorations here relate to this idea that the event of witchcraft or sorcery is both a moment of total uncertainty and emptiness and a moment of constitution. If we can operate with a general definition of witchcraft I would like to add to former definitions this capacity to be constitutive of social relations through its appearing out of non-being.

The Moment of Uncertainty: Witchcraft as Constitutive of Time

There is indeed an important issue of timing in witchcraft. This concerns the process where the diviner is going back and forth between the past and the present in order to establish a temporal scene for the witchcraft event, which draws closer not only the past and the present but also relationships between the witch, the victim and their social environment. We can in general say that witchcraft is an ambiguous and uncertain domain of life. Death by witchcraft is always striking surprisingly and suddenly. It cannot be foretold, it can only be feared. Its origin or cause cannot be clearly deduced, since it does not exactly originate in the agency of one individual person. Generally we can maybe state that witches are creatures that are reported to appear in transformative shape. They are supposedly people who have, consciously or unconsciously, been transformed into transgressive beings that overturn most physical and moral boundaries: they eat human flesh and drink blood, they fly in the air, they take the shape of animals, they become invisible, and they can be in several locations at the same time. They are often associated with fire or light, and are attributed with power and potency. In Balinese Hinduism they are associated with cosmic creative and destructive forces (Stephen 1999), in the sea-dependent Trobriand Islands they are a crucial part of canoe-expeditions (Munn 1986), in Cameroon they are recently implicated in state power and governance regulations (Geschiere 1997). Hence the image of the witch is much more potent that just being a perpetrator of evil, and witches are often central to an imagery of creation and a cosmology of power and constitution (Kapferer 1997). Furthermore, witchcraft indicates a presence that overturns the individual person and makes him or her into something else, an ambiguous being. Reay writes from the Kuma of the Western Highlands of Papua New Guinea:

> Kuma contrast "witches" (*agamp kum*) with "real people" (*agamp wei*). No one is born with a *kum* inside him, but anyone may discover at any age that one has entered his body by jumping from that of a host who has proved unsatisfactory or from a convicted witch at the moment of death. Thus the witch is seen as a humanoid, once fully human, creature, who has lost cortical control and been taken over by a sub-human drive located in his abdomen. (Reay 1976: 2)

Witchcraft is here something that emerges from within the kin group, and that threatens to kill and destroy the group and its very solidarity from within. Hence, the bursting forth of witchcraft lies beyond individual agency and beyond the social fabric, hidden somewhere in the interiors of the person who is ignorant of it, in a spirit double of a person or beyond the personal will in other ambiguous ways.

The work of the diviner is exactly to remove this ambiguity and to nail the event of witchcraft to a time, a place and its personnel. Witchcraft must in this sense be considered as a primary phenomenon; with intentionality and potency to create sociality anew. The witchcraft event, when someone falls over dead from a witchcraft attack, becomes an eclipse of the social relations and their constitution in time and materiality. Its shadow covers temporarily the world as people know it, and it is up to the diviner to reconstitute and retrieve the known world from this ontological darkness.

I have pointed out earlier how the witches or sorcerers on Ambrym Island, Vanuatu, represent a form of totalizing agency that suddenly is seen to intervene in relationships that are marked by jealousy, anger and selfishness (Rio 2002). Of course, terms like "witchcraft" and "sorcery" can only be unsatisfactory and over-coded translations of vernacular concepts that hold somewhat different meanings from our use of these terms in European cultural history. As paralleled in the quote above from Papua New Guinea, what we have to call witch or sorcerer in Vanuatu has a range of different and ambiguous applications. What is called *abiou* in the vernacular of Ambrym Island covers a whole range of phenomena, from simple poisoning with herbs or the use of magic remedies (that we could call "sorcery") to the fantastic creature that is flying through the air and attacking people and eating their intestines (that seems to be more like what we call "the witch"). It is this last capacity of the *abiou* witch that is most seriously feared and that I shall be dealing with here.

The *abiou* reveals itself most potently when someone dies for apparently no reason. They just fall over dead, and this is taken as evidence that they have been killed by *abiou*. Now starts a process where diviners begin to look closer at the circumstances of the death. They search for clues and often locate the death to a time some days before and at a place some distance from the village. They imagine the *abiou* to be a personification of the anger or jealousy that have been displayed by close relatives of the deceased, and it is this personified force that ravages the body of the deceased and eats its intestines. The *abiou* is a creature of the fantastic, a transgressive being who does outrageous, inhuman things to the body of the victim and then stitches him back up and magically brings him back to life so that he can walk around for some days seemingly as normal. This kind of attack is seen to be allowed by the angry relative who has attracted the *abiou* force inside the relationship, even though the responsibility and agency here is completely unclear and misty. A person who is accused of having brought in the *abiou* will be as surprised and uncertain as everyone else.

In this sense the *abiou* intervene in the social relations as third parties who constitute relationships anew by drawing social connections that have not previously been made. Suddenly a father and a son, brothers or cousins become related in new ways through a death that has occurred in their midst. Whereas people thought they had a good relation, they will now discover that their relation was in fact lethally going wrong. It is in this sense that the event of sorcery or witchcraft casts a shadow of uncertainty over the known relations so that the job of the diviners becomes to recast the entire social situation. They will have to go step by step back in time and into the social terrain in order to create and explain a sequence of unknown relationships, allegiances and movements. As Melissa Demian formulates it for sorcery trials in Papua New Guinea: "The history of a death moves 'backwards' through the person of the sorcerer and into his relationship with the family of his victim, their allies and enemies" (Demian 2003: 106).

The *abiou* is no one in particular and potentially each and every one of one's fellow villagers. It is maybe most potently described as a force, although personalized in vivid imagination. It can strike anyone at any time, and since it cannot be foreseen it always only becomes visible after the fact. People know and fear that *abiou* is always looking for relations that have been marked by concealed anger, jealousy or selfishness, and they always try to keep their path clean. In Ambrym it will be liable to strike those who make something of themselves, people who have money, who work, those who hold positions in stores, in church or in politics. These aspects are at least the ones held out *after* sorcery strikes. This victimization of the well-to-do, of the industrial and of the figures of leadership is then superficially creating a reactionary morality—a sense of good versus evil—

where the good is the holding of money, i.e. the right to actually hold money instead of giving, the creation of store-accounts, and further, initiatives of social creations such as community projects of healthcare, church groups, water supply, sports and taking onboard NGOs such as US peace corps volunteers. The warding off of *abiou* is the constant preoccupation of people in these kinds of positions, and its resistance takes place through generosity, walking around to all kinds of kinship ceremonies and church fundraisings and thereby warding off anger and jealousy through a display of openness and generosity.

In this way witchcraft or sorcery is a social force that emerges from the empty space of relations that are in-the-making; in the formation of relational emotions such as anger and jealousy that grow inside people in concealment. At the same time the only proof of concealed anger, jealousy and desire comes through the *abiou* attack. In this way sorcery is in a way the "un-presentation" of relations that present themselves in social situations. Ethnographically this does in some ways seem very similar to the situation described by Evans-Pritchard for the Azande:

> What Azande call *mbisimo mangu*, the soul of witchcraft, is a concept that bridges over the distance between the person of the witch and the person of his victim. Some such explanation is necessary to account for the fact that a witch was in his hut at the time he is supposed to have injured someone. (Evans-Pritchard 1976: 10)

"The soul of witchcraft" among the Azande is hence a connection between persons that has not yet been made and that is maybe not made except through witchcraft. It only presents itself when the witchcraft is a fact, as *having been* the un-presented of the actual relation in question. We have learnt from Evans-Pritchard's ethnography that it could be spotted when flying through the air and emitting a bright light at night, as this represented the only sign of that intentionality that was hidden. The light is the only manifestation that attributes reality to this unrealized intentionality. The only proof of the reality of the un-presented witch is that light escapes it.

On Ambrym Island, the equivalent of the Azande soul of witchcraft would be the essence of the *abiou* creature itself. If a man is angry (*lolfrifri* lit. "inside is hot") at someone and he is keeping this to himself in his hut, the *abiou* will possibly detect it and be attracted into the relation. The *abiou* will use it as an opening and an energy to attack the subject of the anger, without the angry man ever leaving his house. Hence, when people say that they fear *abiou*, they are both expressing the fear that they will be attacked by *abiou* and that they can become responsible for *abiou* attack. This intervention in relationships of a third encompassing party, the materialization of the un-presented social forces themselves in the imagined body of the *abiou*, attacks directly relationships. Most visibly it attacks the victim, of course, strikes him down and eats his intestines, but it also opens up the society in a wider sense to a form of agency that it cannot control.

In the process of court hearings it is the relationship that is now being scrutinized, and the deceased victim leaves behind the problem of what went wrong in the relation. Typically the assumed agent is confronted. Has he been concealing his anger against his brother? Did the relationship suffer because the victim withheld money from him? Why did the brother of the victim withdraw from socializing with him, did he curse his brother from behind his withdrawal? In this way the event of the *abiou* attack and the following village trial have to re-constitute the relationship from that previous relation. The moral agency of thirdness is then expressing itself not on the good qualities of giving and the bad qualities of taking—but works as a direct reaction to social development itself. Taking, killing and eating are as such totalizing acts that set the parameters for social value itself. They both strike down the victim who has been too selfish, too careless with his social obligations, too cocky and "stronghead" as they say in pidgin, *and* the angry, the jealous, the relative who has been letting his anger grow inside his heart. But it more importantly collects the social relation, consumes it and recycles it through the re-evaluation of the social situation that follows.

Abiou thereby creates a total situation of morality—not on the part of the individual person but on the part of the community. It is the community that now has to evaluate how it is developing, how one is on the wrong path, how one has neglected relationships, neglected church meetings, neglected kinship ceremonies, having unmarried couples in the village, having boys uncircumcised in the community, having not assembled enough pigs for mortuary ceremonies, etc. In a most explicit instance of this in the village where I lived on the island of Ambrym, one such sorcery trial resulted in the claim that an *abiou* item had been planted permanently in the village. It would destroy the whole village over some time, and morality became an immediate concern. Discussions roamed between the different churches over who held the key to resist this item, and at one point people were planning to leave. *Abiou* put into perspective the quality of social relations in the village, and the fear of this judgment threatened to break it apart. The witchcraft phenomenon is in this way always hiding itself beyond immediate social reality, and it has to be lured out through divination practices. This process can easily be confused with the investigation of a homicide before a court of law, and this misunderstanding is now also built into state approaches to sorcery (see Rio 2010), but we need to uphold the aspect that divination procedures always take us into another realm of being, a realm beyond being.

Divination on the Margins of Time and Materials

In the following, this chapter will thus mostly be trying to establish a view of that ambiguous space of "not-being" that I think is central to conceptions of witchcraft. This is often the center of attention in divination, and divination techniques are therefore crucial to my argument that witchcraft is a phenomenon of the unknown, of not here and not now, and in this sense a phenomenon of nothingness. Looking back at the last century of theories on witchcraft, it is surprising to see how little attention is paid to the intentionality of divination practices and what that tells us about witchcraft itself. This is also Tedlock's conclusion in her summary of a review of literature on divination:

> Unfortunately, the lack of such ethnography has not stopped scholars from attempting to explain away divination by placing it within evolutionist, diffusionist, ecological, or functionalist theories. All of these approaches rationalise divination after the fact, totally removing it from the realm of intentionally effective action. (Tedlock 2001: 194–5)

This comment encourages us to commence our understanding of witchcraft in the actual formulations of practices of divination and the construction of the alternative reality that they resort to (see also Stroeken 2010). As recently also pointed out by Graw for the case of Senegal, divination carries a lot of constitutional force for the crucial intentionality of a divination space outside of the lived world:

> With the articulation of the *nganiyo*, Senegambian divination becomes immediately performative by opening up what I call an "intentional space"—a performative cultural space for articulating and dealing with personal intentions and desires that not only reflects the intentional nature of the human being but responds to, negotiates and transforms the subject's intentional position by allowing him to engage actively with his longings and afflictions in a changing and challenging contemporary world. (Graw 2006: 81)

There is a very illustrating and ideal example in another article from Senegal (Eglash 1997). Among the Bamana people certain knowledgeable and powerful people can predict futures or trace reasons or origins of misfortune or illness by drawing patterns in the sand. It is a practice referred to as "geomancy" by Graw (2006). Combined with smoking marijuana and drinking palm liquor they tell you how things connect, past or future. When approached by a client the Bamana diviner draws random dotted lines in the sand—in order to deduce geometrical figures as a combination of one or two and two strokes. They are each read as a symbol indicating sentiments or states of being in the client, like for instance "desire." Each figure is put into various "houses"; one of which is called "this world" and another "the other world." The author only makes one example of the meaning of these houses and emotional states: when the figure of "desire" is located in the house of "travel" this says something of the patient's need for travel (Eglash 1997: 117). Based on the total pattern the diviner will deduce the reason behind the client's misgivings as well as the remedy. A more developed analysis of geomantic meaning is given by Graw (2006), as he indicates that the patient more fundamentally now becomes aware of intentionality itself as a precondition for human existence: "as an expression of the ecstatic a priori temporality of the subject" (2006: 114). Through the lines in the sand and the patterns "given" in randomness, the subject becomes directed to the process of time that is crucial for his condition (see also Graw 2009; de Boeck and Devisch 1994).

Through these types of practices we realize that the complex pattern that emerges is itself given the authority to place the client into a total context of sociality—in which his or her life is reinterpreted and reinvigorated with meaning. It also repositions the victim, and his consciousness as pointed out by Graw, so that he also becomes aware of his total situation in a new way. And we note that the meaningful pattern emerges out of nothing—a meaningless, empty and random drawing in the sand. In order to access the pattern itself—and in a sense reaching into that nothingness—the diviner has to have particular powers so that he can enter another state of mind and see things differently. Also on Ambrym Island the diviner is someone with extraordinary capacities of seeing and connecting past, present and future. They are called *vanten ne hal*, which literally means "men of the path" and this connotes their capacity of realizing the sequence of happenings that constitutes the witchcraft attack. They can detail the path for the victim, so that he can intentionally begin to engage with it. Beginning from a clean slate, the diviner amasses a whole system of social meaning for the patient. Thus a concealed, empty space is forced to release meaning and significance, and what was unspoken and incomprehensible is now suddenly exploding into a multitude of meaning and signification.

There are many other examples across the world of similar practices. Marcia Ascher has synthesized such practices of divination from different places (Ascher 2002). As with Eglash she is also preoccupied with the mathematics implied in divination, giving attention to the process of attributing meaning to the combinatory potential of what is given as random patterns—she calls this process "randomizing mechanisms." But I would argue that what is picked up as mathematics in these cases is instead a preoccupation with numbers as an idiom of social ontology: in the sense that divination reveals to the subject his or her total existence through such simple heuristic devises as strokes in the sand. In my argument these types of practices are the opposite of randomizing mechanisms. They are instead constitutional devices that draw a meaningful universe from an assemblage or pattern that emerges out of that very significant sphere that is not-being.

Typically, when diviners want to draw out some truth from a hidden reality this is often done through killing some small edible animal, such as a crab, a chicken or a snake. Inside its vital organs lie the clues to the interpretation that is the divinatory work. The truth from *beyond life* is often accessed by attention to essential life substances such as blood, grease or intestines. Koen Stroeken has recently given a description of how this works among the Sukuma of Tanzania (Stroeken 2010: 152–83). Here the opened chicken gives a window into an unknown world, and inside the intestines the oracle merges the intentionalities, or "paths" in the language of Ambrym divination, of the victim, the diviner, the witch and the ancestors so that they all become available and within reach to each other.

But the variation in the technologies of diviners is enormous, and the case seems to be that every society described has its own way of teasing out proof and indication of the circumstances of witchcraft or sorcery; of whom, when and how.

In Evans-Pritchard's account of the Azande they either give the oracle voice through poisoning a chicken and asking it for names, or the voice is given to objects that are induced to move in particular directions in order to point out the origin of the malevolence, the main example being the "rubbing board" (Evans-Pritchard 1976). Another classical example, now in the region of Melanesia, is the account of Trobriand sorcery by Reo Fortune (1932). Here divination was usually done by "water-gazing":

> He (the diviner) cuts the water-in-changed-nature open. At the bottom of the cut he sees the spirit of the witch who has abstracted the spirit of his patient and who now has it concealed, or the spirit of the sorcerer who has the *sumwana*, body leavings of the patient and now has them concealed. (1932: 154)

In another example from Melanesia, from the archipelago next to the Trobriand Islands, F.L.S. Bell describes the divination rite among the Tanga people of New Ireland, Papua New Guinea. This was reported in 1935 to the Royal Anthropological Institute. A case is followed where relatives of a deceased girl use a method called "fishing for the ghost" in "an attempt to discover the identity of the sorcerer responsible for the death" (Bell 1935: 84). They take a long bamboo pole and attach a bundle of scented leaves and flesh from a pig to the end of it. With the pole on their shoulder they walk from the grave of the girl as they try to entice her ghost into coming with them, tempting her by the smell of the leaves and the cooked meat, on an expedition into the bush and back to the village. Ensuring that they have managed to return the ghost to the village, they line up in order to question the ghost—to give them some answers in return for the gift of meat. They ask "who has made this evil sorcery against you?" (1935: 85), and they suggest name after name and wait for the pole to move. In this moment they are hence searching for answers in the social terrain of the girl's relations. As they say a particular name the pole starts to shake at the end, evidence that the ghost is trying to snap the whole bait of pig meat from them along with this name of the true sorcerer. After this evidence was out in the open, Bell comments, no punitive action existed, but the sorcerer would be "publicly convicted as an unsocial being," i.e. cast out of social life in the village (1935: 86). By divination they endeavor to transform the unknown into the known, the unnamed into the named. They try to overcome the "why" with a "who" by performing this ritual. They take one step back in time by awakening the dead girl, in order to reveal sorcery itself in its own realm and in the very event of taking place. They reach into the realm of the dead spirits and call into this ambiguous space for some answers. The truth of the event does not, at first, lie within the realm of the living, it lies hidden inside the spiritual realm. The circumstances of the sorcery death must be extracted from this realm before it can become real, hence the institution of the ritual of "fishing" into uncertainty itself.

In yet other circumstances it is the accused person himself or herself who is forced to divine the circumstances of the witchcraft. Levi-Strauss has used an example from the Zuni of New Mexico to look at the constitutive process of a sorcery trial (Levi-Strauss 1963: 174). In this case a boy is accused of having ensorcelled a young girl. At first he denies this, but as the divination process goes on he begins to confess. He becomes a confessional voice that retrieves information beyond his own consciousness about his mischievous activities. In Levi-Strauss' view he is thus starting to fantasize about himself as a sorcerer, and thus becomes one. We can see the similarity to Bell's case from Tanga Island. The construction of truth takes place through the medium that is chosen to speak. Whereas the ghost of the dead girl, speaking through shaking the pole with meat, is forced to give her report in the first case, the case of the confessing Zuni boy is more straightforward as it comes directly from the boy himself. It has the same character, as the audience trusts him as an agent to mediate the truth from a realm that is until then uncertain. It is the self-producing narration of the witchcraft institution that structures the witchcraft event—as a truth beyond doubt (see also Holbraad 2012).

In the contemporary situation in Vanuatu that I have experienced in recent fieldwork the beating up or even killing of accused witches has some of the same effect (see Rio 2010, 2011). Normally small incidents of bad luck or illness calls for the visit to a diviner, or a "clever" as they call them in Vanuatu Pidgin English. These diviners in Vanuatu use no particular tools for divination; they look into water, they blow into the ear of their patient, or they just "see" both past and future episodes in the life of the patient. In the case of homicidal witchcraft they inspect the victim, as a forensic examination, and search for clues in the skin of the victim, his bodily composition at death, angle of the tongue, etc. They thus come to "see" a totality of relationships and events, as a reconstruction of the "path" of the last moments in the life of the victim, and this leads them to direct attention to particular men who are accused of causing the misfortune. As explained above this is not about the accused actually committing the outrageous act, but more a matter of "calling it forth" by concealing anger or jealousy. The further treatment of the accused seems to indicate a process of divination continuing into the trial. Here the accused is subject to a bodily attack, and he is forced to talk about the killing. The diviner is not active in this process, but his testimony or analysis is taken as evidence. The crowd now envelopes the divination medium, the victim and the unknown realm of witchcraft into one single site—in the body of the accused—and this becomes a scene of violence and divination simultaneously. They are collecting clues for a while, and they thereby synthesize all signs and evidence into a virtual narrative about events, people who have been moving around in the landscape and leaving certain traces, and further clues like traces of blood or otherwise unusual circumstances. It is like a police murder investigation, but inbuilt with a fantasizing narrative of people flying in the air, transforming themselves into other people or animals, being in different places at the same time and as having motivations of greed and desires that you don't see in people's daily lives.

In their direct act of accusation and violence the crowd is throwing out accusations about these fantastic events, and at the same time they are deforming the accused into a bloody, unrecognizable creature and thus trying to draw out some words of confession from his new state of being. They beat him up in order to break through to a different form of agency—one that lies beyond his conscious person. They in a sense try to get access back to the *abiou* mediating creature by again going through the accused—just like the *abiou* went through him to get access to the victim. The deconstruction of the known person of the accused into a pure sorcerer's voice constructs him as a mediator of a hidden truth. In these courts in Vanuatu the accused often do confess to their own great surprise. They then also recognize afterwards that they could have done it without being aware of it—but under the influence of a destabilizing agency such as the *abiou*. In many ways this is not seen as entirely their individual responsibility, and even though the accused will be often be a suspect for possibly mediating malevolence and destruction again, they go back to a normal life with their relatives and no consequence is then made of the fantastic narrative of the witch trials.

In Levi-Strauss' framework this forced narrative structure of divination is taken as indication that the event of sorcery or witchcraft itself is non-existing. The assumption of Levi-Strauss, supported by Evans-Pritchard, is that "witchcraft has no real existence" (Evans-Pritchard 1976: 43). We learnt from Evans-Pritchard in the case of the Azande that the phenomenon itself only starts to gain reality when signs start emanating from the mediation of the rubbing-board or poisoned chicken through the acts of the oracle. These authors thereby manage to maneuver around their problem with the apparent irrationality of witchcraft. Sorcery or witchcraft is rational because it is an effect of language and narration. It attributes meaning and reality to what is unknown in society.

I believe these analyses touch on an important point about divination: the understanding of death and illness requires contact with a realm of the social that is not immediately reachable within our immediate social relations. It represents a challenge from a realm that is not entirely existing, at least un-existing in terms of immediate, visible social relations and of language, and it forces us to question that which is existing, to look at it from another perspective. But this does not need to imply that witchcraft has no existence; simply that the existence is of the unknown or a potentiality of the unknown.

My first point, then, is to address the phenomenology of these various attempts at divination. What kind of a practice is divination and what does divination tell us about witchcraft? Before witchcraft can be uncovered, divination has first to take one step back in time, in order to reconstruct as space for the understanding of the unique circumstances of the witch attack, and divination is for the analysis the only entrance into witchcraft also. For the purpose of understanding this process on its own terms Bruce Kapferer has introduced us to the idea that sorcery divination in Sri Lanka is taking place in the ritual circumstance of a "virtual space":

> a dynamic that allows for all kinds of potentialities of human experience to take shape and form. It is, in effect, a self-contained imaginal space – at once a construction but a construction that enables participants to break free from the constraints of determinations of everyday life and even from the determinations of the constructed ritual virtual space itself … The phantasmagoric space of ritual virtuality may be conceived not only as a space whose dynamic interrupts prior determining processes but also as a space in which participants can

> reimagine (and redirect or reorient themselves) into the everyday circumstances of life. (Kapferer 2004: 45)

The arena of divination becomes a space that is structured outside of the real world, yet usefully relating itself to all the known elements of the real world and the elements of the misgivings affecting the victim of witchcraft. In much the same terms as Kapferer, Koen Stroken proposes that divination among the Sukuma of Tanzania offers a therapeutic restructuring of the world. Inside the microcosmic world of chicken entrails the oracle resides and "gives" the diviner the world as seen by the ancestors (Stroeken 2010: 181). A similar vision of so-called basket-divination among Ndembu has also been reported by de Boeck and Devisch in their revision of Turner's ritual analysis: "Seen in its own terms, divination does not so much offer a mimetic model of a social context, *but rather makes a world*" (de Boeck and Devisch 1994: 99, emphasis added). In all these analyses there is an effort of a social environment to become communally focused or directed toward the moment of death of one of its members, as a space of particularly moral and sociological potency. The search necessarily has to reach into a realm where things are uncertain and partly unknown. It is a space where the known social circumstances of a forensic nature become twisted, doubted and perverted, and the process of divination and witch-hunt can again bring certainty to this domain in a process of "world-making." The world that is constructed in divination is then fundamentally and paradoxically steering the intentionality of the affected parties to an un-presented world. The remedies and the changes in consciousness of the victim and the accused witch are all part of this un-presented world and have to remain there. The world presented by the diviner, a "phantasmagoric" world in the vocabulary of Kapferer, is never meant to enter or mix with the everyday world, but for a suspended time it *is* their world. This is the fact established by divination, brought on by the illness or death as equally phantasmagoric phenomena and evidence of an opening of not-being into being. What divination tells us is always that witchcraft is the alternative reality of the phenomenon in question. On Ambrym this alternative reality is that presented by the personified creature called the *abiou*—who potentially stands inside every social relationship and threatens to destroy it. The thin veneer that separates these domains is potently hidden in the most common of objects, in a materiality of next to nothing: in water poured into a bowl, the intestines of a chicken, in dotted lines in the sand or mumbling sounds from a bloody accused, because these objects next to nothing can open windows into nothingness itself as a domain of potency and force.

Further, following from this, the remedy for what one sees in the vision of the diviner is in the end the concealment of this vision, since the opening of not-being into being, or the vision of the *abiou* into the social relations in the Ambrym context, is an unbearable crisis, however thin that veneer is. The remedy in Vanuatu is the beating and silencing of the accused sorcerer, or even killing of the witch as in recent events (see Rio 2010, 2011; see also Knauft 1985).

The Not-Yet-Being of Witchcraft: Accounting for the Void

Following Evans-Pritchard and Levi-Strauss, a general trend in the understanding of witchcraft is the idea that the word is ultimately the key to understanding the power of both witch and witch-hunter (Favret-Saada 1980; Siegel 2003, 2006; West 2007). In the moment when someone says "he is a witch," this is already creating new social circumstances. Witchcraft is supposedly formulated as an after-the-fact speech act that transforms the real world. The referents of the signs expressed have no reality outside of language. The anthropologist is ready to accept sorcery or witchcraft as a sociological given only as far as the "as if" becomes "is" through language (West 2007: 64).

These authors all take for granted that only the phenomenon that finds its shape in discourse is the real phenomenon—in contrast to that which is before the reality of discourse. This still means that reality commences when the rubbing-board speaks. This is, however, in obvious conflict with how people live with sorcery or witchcraft. They clearly perceive death or illness itself as a major and troublesome part of their reality, and when they act on it, this is what they act on and what they start engaging with in discourse. My claim is that they always act in reaction to or even in resistance to what the event itself tells them. Whereas these authors have been caught in the idea that the witch-hunt creates witchcraft as its language, my claim is that witchcraft expresses a certain form of consciousness that brings about the witch-hunt as its resistance. This is taking place after divination has opened up the world to a different perception of it. This claim thus depends on the willingness to allow ontological status to the moment of witchcraft itself, and to allow reality to something that is not existing—yet—through the divination procedure. The intense activity of naming and constructive discourse of guilt and accusation is as I see it exactly the activity of *hiding* the fact that one conceives of this non-being as a being. What comes after divination is not its continuation, of trying to understand the world as seen in the intestines or cut water of the diviner, it is rather a break and a denial of that perception. It is exactly the world of divination that one does not want. The crucial moral point about the witchcraft event is that it threatens to overturn reality, as I noted above, by eclipsing reality temporarily. Divination is where language starts as an effort, first to lay out all the elements of this crisis, and further, to deny that potentiality.

In Kapferer's analysis of Sri Lanka Suniyama rite, this curing rite is an effort to create a virtual reality: "The Suniyama is a virtuality in the sense that it is an organization of activities that are integral to the routine activities in the lived-in life world but not subject to the indeterminacies of its processes" (Kapferer 1997: 179). The rite constructs a world around the sorcery victim where the forces affecting the victim can be controlled. It is stopping time, and creating a timeless space for replaying the negative influences affecting the victim. It is a virtual world created for the purpose of removing that potentiality of uncertainty or nothingness that the illness has revealed. The ritual space is set up to control that space that is beyond control and hidden from view in normal social life, and to purposefully treat the victim, *while being in that space of sorcery itself.* In that moment, Kapferer adds, "the rite does not simply dissolve into reality but brings it forth" (1997: 180). This is altogether a more radical point than saying that discourse and naming is the foundation of sorcery or that language of sorcery transforms the "as if" into an "is." The Suniyama ritual itself is actually a way of explaining that sorcery is its own constitutional realm, and is trying to reconstruct that constitutional realm as a stage for the victim to be healed inside. This ritual is one solution to the unbearable problem that sorcery brings about; that of the void or nothingness. In the other accounts of divination referred to above from Africa or Melanesia, there are other ways of acting on the world opened up by divination. For the case of Vanuatu I have tried to argue that beating up the witch is another, more paradoxical, way of accessing and closing at the same time the world of divination. However, in all the examples we need to account for the world of sorcery or witchcraft, glimpsed in a language of divination, as a real category of life.

Like Kapferer, but coming from the angle of logic and physics, the philosopher Alain Badiou has also tried to establish a positive theory about the existence of that which is not, the void. It is of interest to note the way he establishes the void as a necessary ontological existence for the language of mathematics. To the degree that mathematics (of divination or statistics for example) is always only trying to be a tool for understanding and accounting for social reality, this should also be of relevance for anthropology. Keeping in mind the above discussion of a situation of witchcraft as an indeterminate situation of a determinate social environment, we can maybe relate to the following:

> By itself, the nothing is no more than the name of unpresentation in presentation. Its status of being results from the following: one has to admit that if the one results, then "something" – which is not an in-situation-term, and which is thus nothing – has not been counted, this "something" being that it was necessary that the operation of the count-as-one operate. Thus it comes down to exactly the same thing to say that the nothing is the operation of the count – which, as source of the one, is not itself counted – and to say that the nothing is the pure multiple upon which the count operates – which "in-itself", as non-counted, is quite distinct from how it turns out according to the count. (2006: 55)
>
> Ontology can only count the void as existent. (2006: 58)

For me this ability of a logician to allow for the ontological determinancy of "that which is not" over "that which is" offers a critique of the anthropological fear of accounting for the void of social life. I perceive Badiou's reflection to be on the same level as the realization that witchcraft is a particular form of social existence—that "the nothing is the operation of the count." In the history of numbers and mathematics, it should not be forgotten that in medieval Europe zero became the *diabolon* set up in contrast to the *symbolon*, as a relation of chaos to cosmos, when the status of numbers and mathematics was not yet set apart from social values, or indeed, divination practices (see Herrenschmidt 2007). In that world the zero was perceived as the devil, a destructive force closely associated with witchcraft itself. That was also why Gerbert d'Aurillac, later to become Pope Sylvester II, left out the zero when he introduced Indo-Arabic numbers to European mathematics at the turn of the first millennium. Due to his involvement with numbers, despite this omission and possibly due to his keeping the zero to himself, he became known as the "Sorcerer Pope" and after his death his bones were dug out of his tomb (Flusche 2005).

This deeply cosmological engagement in mathematics with the non-number zero, fearing the void as possibly a destructive force when accounting for the world in real numbers, can be compared to how people who believe in witchcraft acknowledge that any appearance of social relations or countable things, persons, statuses, etc., must also have its own lurking presence of un-appearance. After all, all things come into being, and in that moment or space of becoming, something must remain, at least as a potentiality. That is the void of social reality, which people see as constantly bringing into existence new things, but also bearing a possibility of destruction.

Thus we can begin to allow people the right to *think* witchcraft, and not only *speak* it. Once we grant them this capability, ethnography can provide us with the crucial materials for realizing how people operate and struggle with the ontological structures of their reality.

Conclusion: The Gift of Death

I have earlier argued with reference to Mauss' gift that the people of Ambrym, Vanuatu, do not return gifts in the sense that there is a continuity or similarity between gift and return-gift (Rio 2007). I describe how people in kinship ceremonies kill pigs and offer the meat to the people who have provided them with the "gift of life," i.e. the blood, nurture and clothing coming from the mother's kin. I have insisted that these pigs are killed as a *denial* of that gift, and hence killing the spirit of the gift of life. The gift of life is not what people want. They want to be self-contained and so they turn the hierarchical perspective of the gift into a perspective of reciprocity as balanced, equal relationships. Reciprocity dissolves the totalizing perspective in the first gift; the gift that gives them life. For people on Ambrym this is an intuitive reaction to the kind of agency that is seen to be totalizing. The risk is always that one may become totally overtaken by that sort of perspective that one is not in charge of. When young boys kill pigs for their mother's kin in circumcision ceremonies it implies the end of *given* life and motherly influence and the start of reciprocity with their relatives. Analogically, I would argue that in witchcraft also, relations cannot be assumed to be given in their dyadic structure of intrigue. It could be that the structure that we see appearing after the fact—as a structure of suspect and victim—is instead the reaction of people to what they perceive to be a totalizing event: not the gift of life this time, but the gift of death.

This is also something they do not want—and so they turn it into something else. What the anthropologist now sees before him are the proceedings of the witch-hunt. But what we then miss is the understanding of the temporal logic of the witchcraft attack itself that is now negated. The witch-hunt is getting back into order a process that is off joint.

The response to the constitutional event that brings evidence of that which is not known, is desperately to turn it into something which is or that can become known—i.e. to search for explanations, to search for a guilty party and to reciprocate and avenge. What we see before us are the dyadic relations in the attack of the witch-hunt in compensation claims, and in the state apparatuses we see searches for just cause, intentional homicide and scene of the crime. However, inside a logic of witchcraft what is received in the event is a perspective on social relations from the un-becoming, a judgment seen to be intervening in these dyadic relations. In the denial of this totalizing perspective from thirdness there is a manic effort to create sorcery as individual agency. Mobs rage across the village or suburb in order to find the agent—and they force a change of the totalizing perspective of the witch attack by instead locating cause and effect in their dealing with the suspected murderer. Beating up and killing the sorcerer is the direct return and negation of this gift of death.

In the suburbs of Port Vila there are often now such raids. Mobs of young and aggressive people drive out a suspect. In the unstoppable rage and violence against the accused—in despair over a deceased relative—they press the suspect into confessing. Most often guilt is left open to be speculated upon, but these events of both divination and punishment manage to turn the perspective of witchcraft around. The reaction to this total situation of crisis brought on by the witchcraft event is to seek out a second party in order to shift the situation of totality, a situation where the negative space of a void unfolds itself as reality, into a logic of accountability, reciprocity and transaction. I suggest that the aggression and desire to establish guilt must represent emotions that cannot find expressions in other ways. I have hinted that the beating and deformation of the accused suggests that the mob seeks beyond his immediate personality or self. They look for answers that he himself cannot express, and in the same instance as they try to beat answers out of him they instead silence him. The silencing or even the killing of the accused carries forth its own imperfect solution to an unbearable problem: the problem that forces beyond control have turned up in the midst of one's social relations. Guilt has to be established as a closure and silencing of the truth of the underlying social circumstances.

Witchcraft always operates with its own language, outside language. This is because witchcraft as an encompassing phenomenon is capable of taking in and totalizing a wide range of issues in a society. But witchcraft always comes first—in a fundamental way. It has itself a constitutional mode to it as it causes a fundamental scrutiny of the social relations and conditions that it eclipses. It can never be foretold, it always strikes unexpectedly, and each strike always constitutes relations anew. Such is the technology of death.

References

Ascher, Marcia. 2002. *Mathematics Elsewhere: An Exploration of Ideas across Cultures*. Princeton; Princeton University Press.

Badiou, A. 2006. *Being and Event*. London: Continuum Books.

Bell, F.L.S. 1935. The Divination of Sorcery in Melanesia, *Man* 35(2): 84–6.

de Boeck, F. and R. Devisch 1994. Ndembu, Luunda and Yaka Divination Compared: From Representation and Social Engineering to Embodiment and Worldmaking, *Journal of Religion in Africa* 24(2): 98–133.

Demian, Melissa 2003. Custom in the Courtroom, Law in the Village: Legal Transformations in Papua New Guinea, *Journal of the Royal Anthropological Institute* 9: 97–115.

Eglash, R. 1997. Bamana Sand Divination: Recursion in Ethnomathematics, *American Anthropologist* 99(1): 112–22.

Evans-Pritchard, E.E. 1976 (1937). *Witchcraft, Oracles and Magic among the Azande of the Anglo-Egyptian Sudan*. Oxford: Clarendon Press.

Favret-Saada, J. 1980. *Deadly Words: Witchcraft in the Bocage*. Cambridge: Cambridge University Press.

Flusche, A.M. 2005. *The Life and Legend of Gerbert of Aurillac: The Organ Builder Who Became Pope Sylvester II*. Lewiston: Edwin Mellen Press.

Fortune, R.F. 1932. *Sorcerers of Dobu*. London: Routledge & Kegan Paul.

Geschiere, P. 1997. *The Modernity of Witchcraft: Politics and the Occult in Postcolonial Africa*. Charlottesville: University of Virginia Press.

Graw, K. 2006. Locating *Nganiyo*: Divination as Intentional Space, *Journal of Religion in Africa* 36(1): 78–119.

Graw, K. 2009. Beyond Expertise: Specialist Agency and the Autonomy of the Divinatory Ritual Process, *Africa* 79(1): 92–109.

Herrenschmidt, C. 2007. *Les Trois Écritures: Langue, Nombre, Code*. Paris: Gallimard.

Holbraad, Martin. 2012. Truth Beyond Doubt: Ifá Oracles in Havana, *HAU* 1(2).

Kapferer, Bruce. 1997. *The Feast of the Sorcerer: Practices of Consciousness and Power*. Chicago: Chicago University Press.

Kapferer, Bruce. 2004. Ritual Dynamics and Virtual Practice: Beyond Representation and Meaning, *Social Analysis* 48(2): 33–54.

Knauft, Bruce. 1985. *Good Company and Violence: Sorcery and Social Action in a Lowland New Guinea Society*. Berkeley: University of California Press.

Levi-Strauss, C. 1963. *Structural Anthropology, Vol. I*. New York: Basic Books.

Munn, N. 1986. *The Fame of Gawa*. Durham, NC: Duke University Press.

Reay, M. 1976. The Politics of a Witch Killing, *Oceania* 47(1): 1–20.

Rio, Knut. 2002. The Sorcerer as an Absented Third Person: Formations of Fear and Anger in Vanuatu, in *Beyond Rationalism. Rethinking Magic, Whitchcraft and Sorcery*, edited by B. Kapferer, pp. 129–54. Oxford: Berghahn Books.

Rio, Knut. 2007. Denying the Gift: Aspects of the Gift and its Counter-Prestation in a Vanuatu Society, *Anthropological Theory* 7(4): 449–70.

Rio, Knut. 2010. Handling Sorcery in a State System of Law: Magic, Violence and Kastom in Vanuatu, *Oceania* 80(2): 183–97.
Rio, Knut. 2011. Policing the Holy Nation: The State and Righteous Violence in Vanuatu, *Oceania* 81(1): 51–72.
Siegel, J. 2003. The Truth of Sorcery, *Cultural Anthropology* 18(2): 135–55.
Siegel, J. 2006. *Naming the Witch*. Stanford: Stanford University Press.
Stephen, M. 1999. Witchcraft, Grief and the Ambivalence of Emotions, *American Ethnologist* 26(3): 711–37.
Stroeken, K. 2010. *Moral Power: The Magic of Witchcraft*. Oxford: Berghahn Books.
Tedlock, B. 2001. Divination as a Way of Knowing: Embodiment, Visualisation, Narrative, and Interpretation, *Folklore* 112(2): 189–97.
West, H. 2007. *Ethnographic Sorcery*. Chicago: University of Chicago Press.

Chapter 3
Time, Late Modernity and the Demise of Forever: From Eternal Salvation to Completed Bucket Lists

Michael C. Kearl and Michael Hviid Jacobsen

Introduction

Philosophers ask whether individuals can live without confidence in something indestructible within themselves. Perhaps it is because we are the only species aware of our own eventual deaths that the most mystical of experiences involve time, particularly when time slows down or even stops. The experience may last but an instant but its 'after-glow' lasts for days (Orme 1962/9: 23). No time, no death. As Darryl Reanney aptly observed in his book *The Death of Forever*, "death and time are indissolubly linked" (Reanney 1991: 19), because man is the only living creature aware of time and its passing.

The awareness of time—its stopping, passing, ticking or slipping away—is also an awareness of death. Most often, individuals suspend this awareness and live their lives as if they were forever. This suspension is indispensable in order to live life and create any meaning with the absurdity of living life toward death. As Erving Goffman once mused:

> Poets and the religious are wont to argue that if an individual compares the very considerable time he is slated to spend dead with the relatively brief time allowed him to strut and fret in this world, he might well find reason for viewing all of his life as a very fateful play of very short span, every second of which should fill him with anxiety about what is used up. And in truth, our rather brief time *is* ticking away, but we seem only to hold our breath for seconds and minutes of it. (Goffman 1967: 261)

Time, and death, is thus the problem of the living. As Norbert Elias once poignantly observed, "death is the problem of the living. Dead people have no problems. Of the many creatures on this earth that die, it is human beings alone for whom dying is a problem … They alone of all living beings *know* that they shall die; they alone can anticipate their own end" (Elias 1985: 3). And as Zygmunt Bauman later added, human beings are the only creatures who "know they know" that they shall die (Bauman 1992a: 1). It is this dreadful knowledge of living a life with death as a definitive outcome that torments the thoughts of human beings.

With death and time so intimately intertwined, the human species has for millennia struggled to find a meaning for the timely nature of life and its eventual dissolution in death. Throughout human history, few thoughts have been so powerful and persistent as thoughts of a postlife existence, thoughts of "something" beyond death. However, the human imagery of immortality, eternity, survival and longevity is as varied as it is ever-changing (Bauman 1992a; Hick 1976; Perrett 1986; Prioreschi 1990; Toynbee 1980). Just as humans constantly work at constructing, deconstructing or carving a meaning out of human life (e.g. society, culture, institutions, habits, norms, etc.), they also constantly and incessantly engage in constructing and deconstructing a meaning with their deaths (Bauman 1992b; Jacobsen 2012).

There is thus something about the related notions of permanence and timelessness that have long appealed to the human imagination. Timelessness is changelessness, the eternal. To defy the changes wrought by time, those of the Neolithic era created huge megaliths in both the Old and New Worlds, spaces "where people felt united with their ancestors, the gods, the earth, and the heavens; a place where participants felt they could transcend time" (Falk 2008: 26). Such endeavors continue. To demonstrate the sacredness of its elite, the Catholic Church publicly displays the incorruptible bodies of its saints. The good do not rot; they remain in a continuous present. The idea proved irresistible to even the godless regimes of Communism. Preserved under glass for public display are the remains of Vladimir Lenin, Ho Chi Minh and Mao Tse-Tung.

But, as Zygmunt Bauman (1992a) notes, contemporary late modern society has fundamentally challenged the timeless paradigm upon which traditional death transcending strategies were constructed. We have learned that death is the fate of not just individual life forms but of entire species and genera, and that the collapse of human civilizations is inevitable as well. We have learned that the passage from existence to non-existence is not simply the fate of just life, but also of mountains, planets, stars, galaxies and, according to the Big Bang eschatology, of eventually the entire universe.

Here we explore some of these contradictory theses of the human condition and the contemporary strategies for encountering and countering our awareness of living on finite time. We begin with the interplay between our species' temporal essence and its obsession with finitude. Next considered are the historically novel temporalities of late modern selves—specifically, dramatically increasing life expectancies and prolongations of the dying process—and of the socio-cultural orders wherein they exist—orders featuring accelerating rates of change and obsolescence, diminishing connections and relevancies of the past, and obsessions with ever-increasing efficiencies with attention focused on ever-diminishing units of time. Finally, we turn to the new *ars moriendi* and how it shifts the temporal themes of mortuary ritual from the future to the past, from reassurances of eternal life to restoring the past lives of the deceased.

The Temporal Primate

The 1996 *Official Statement of Churches Together in England on the Millennium Celebrations* observes how "there is something within the human spirit that needs to mark the passage of time." Experiences of its passage shape our dreams, memories, anxieties, music, pay checks, identities and relationships. Those either who do not sense or who reject their social and cultural rhythms are often adjudged as deviant, people who "march to a different drummer."

There is no question that this preoccupation is part of the British spirit, where Greenwich Mean Time (GMT) begins at 0° longitude and whose Salisbury Cathedral has 12 doors to represent each month of the year, 365 windows for the annual number of days, 60 sculptures for the number of minutes in an hour, and 60 crosses for the number of seconds in a minute. Architectural evidence of this temporal preoccupation dates back to the Stone Age. In Ireland is the New Grange tomb, the oldest known building in the world. This structure, graced with gleaming quartz, was constructed over five centuries before the great Egyptian pyramids—a period of time equivalent to that between Columbus's New World Voyages and flights of the space shuttle. There is a small slit above the entryway that only during dawn of the winter solstice allows light to reach the inner chamber, wherein interred are the remains of four adults and one child. Many generations have speculated on why all the energies were focused on constructing such a huge structure (which took an estimated 20 years) designed to capture but a brief moment in time. Perhaps only at this sacred time—the annual moment when the life-giving forces of sunlight have reached a nadir and will once again regenerate over the span of six lunar cycles—did they believe that the spirits of the dead could be animated.

But those of the British Isles have no monopoly on such temporal acuities in architectural design. Archaeologists have determined that many of the great monoliths of prehistory are but huge chronometers, used to synchronize the rhythms of social life with those of the cosmos. Temporal acknowledgments are to be found in more recent designs, such as the Capitol Building in Washington, DC, which has 365 steps. Further, the temporal imagination is hardly confined to physical structures.

The temporal primate is also obsessed with how things began and how they will end. We are unique in our endeavors to identify our origins and ultimate demise, producing mythologies and sciences about the beginnings and endings of individual lives, species' existence, the planet upon which we reside, and the entire cosmos. We have sent messages into deep space knowing that they will not arrive to any life form capable of deciphering them until we are long extinct. Symbolizing the epitome of human temporal creativity is Albert Einstein's paradigm-busting insight into time's relativity to gravitational mass and speed, and how the fabric of space-time will eventually be filled with black holes.

Evolutionary psychologists are closing in on the survival benefits and reproductive successes associated with humanity's incredible temporal senses. Developmentally, time is one of the later concepts learned and one of the first to be distorted with the dementias of old age. Why are there no old animals in nature and yet a growing number to be found among human primates? There must be some evolutionary advantage to being able to think outside of the present, to comprehend the ideas of past and future, and to respect the prophecies of the longest-lived and longest-remembered. It is, according to the dominant paradigm of thanatologists, the motivating insight underlying our species' creation of civilization.

Unique in the awareness of its inevitable mortality, the human primate understands time to be the ultimate scarcity, making its finiteness a key motivator of social endeavors. "The meaning of life," observed Franz Kafka, "is that it ends." Culture itself, according to Ernest Becker (1973), is the consequence of humanity's longing to transcend mortality and boundary. To avoid death the body self must be denied in favor of a symbolic identity, which may be either individual or collective in form. The objects of such transference (e.g. whether God, nation or tribe) can become the sources of personal enslavement and the root of social evil, as when the illusion of a death-transcending belief becomes confused with reality (think of radical Islam's suicide bomber believing in personal martyrdom and eternity with the 72 virgins) (Becker 1975).

Connections Between Time and Death Understandings

There is a potent force associated with temporal limits, one harnessed by the modern concept of "deadlines," which warps the experiences of lived and future time just as black holes warp and compress the fabric of space-time. It is a force that demands culmination and that brings revelation. Attention given to those at death's door, such as the Randy Pausch "Last Lecture" phenomenon a few years back, tap into the mythic archetypal belief that special ability and insight comes to those at life's end—supposedly the only occasion when individuals feel compelled to derive a sense of coherency and meaningfulness from their encounters with time. At this juncture, when the past washes over the future and when one accepts that one's time has come, appear the basic truths of existence stripped of the banalities that so often preoccupy the living. How else can we explain the numerous "famous last words" compendiums or the deathbed confessions of murderers? In fact, the very word for the ultimate of endings, *apocalypse*, comes from the Greek word meaning "revelation."

The awareness of death has a bearing on conceptions of our place in time. Consider the contemporary attention given to the "last ofs." Not many birds receive newspaper obituaries nor receive a National Historic Landmark. But Martha did. Her death in September 1914 marked the extinction of the passenger pigeon, a creature that less than one century earlier could arguably have been described as the most successful bird species in the United States. Audubon described a flock he observed in the autumn of 1813, so thick the "light of noonday was obscured as by an eclipse." The flyover lasted three days. The death of an individual is one matter, but the approaching death of the last surviving member of some social category, whether the extinction of a species or the perishing of the last survivor of some broadly remembered historical period, is something totally different.

News was made in 2009 when Britain's last survivor of World War I trench warfare, Harry Patch, passed away at the age of 111. Observed British Prime Minister Gordon Brown, "the noblest of all the generations has left us, but they will never be forgotten. We say today with still greater force. We Will Remember Them." The following year Canada's only surviving veteran of the same conflict died, the last of his country's 650,000 who served. Canada's Prime Minister noted how John Babcock "was Canada's last living link to the Great War, which in so many ways marked our coming of age as a nation." But the fiftieth anniversary of World War I did not receive during the 1960s the attention given to the same anniversary of World War II during the 1990s. Countering trends yielding historical ignorance and sense of disconnect with the past were the personal identifications with particular *generations*, a temporal concept that not only gives individuals a sense of place within the timeline of history but also a renewed sense of connection with those long dead and those yet to be born. When, for instance, Mary Josephine Ray, the New Hampshire woman certified as the oldest person living in the United States, died in 2010 at the age of nearly 115, the living had apparently lost connection with one who was alive with people born in the eighteenth century. However, in 2012, came the story of how John Tyler—the tenth President of the United States, a man 19 years older than Abraham Lincoln and one nine years of age when George Washington died—still had two living grandsons.

The African Kiswahili conceptualize the past in terms of such connections and have labels for these different intermediaries: there are the *sasha*, the living dead, the Harry Patches and John Babcocks, whose lives intersected our own. Then there are the *zamani*, the truly dead, whose existence pre-date the lives of society's most superannuated. As will be developed, in exchange for the non-verifiable immortality promises of the past, new technologies now provide a tangible experience of timelessness by making part of the present the images and voices of the *zamani*. Though we act as though we are immortal, at some deep level we increasingly understand that such permanence is an illusion, just as is the "permanent" of a hairdresser.

The awareness of time also has a bearing on death. How small are the odds that three of the 56 signers of the American Declaration of Independence would die by chance on July 4—two of whom, Thomas Jefferson and John Adams, perishing on that day precisely 50 years later? Perhaps one of the most striking findings concerning death's timing was detected by David Phillips and Kenneth A. Feldman (1973), who found by chance that death could be postponed for social reasons. Wanting to demonstrate equiprobability in a statistics course, Phillips had his students plot the month of death of more than 1,300 famous deceased Americans, expecting that there would be found roughly 110 deaths per month. But such was not the case. Instead, they replotted the data in terms of individuals' month of birth and examined monthly death rates before and after anniversary months. They found a relationship between birthday and death months, with lower than statistically expected death rates during the month prior to the birth month (a "death dip") and greater than statistically expected rates for the month immediately following (a "death peak"). Phillips (1972) postulated that famous persons delay their time of death in order to receive the public attention and expressions of respect that typically come with their birthdays.

The "Death System" of Late Modern Societies

If one believes that one key to understanding cultural systems is to examine their so-called "death systems" (Kastenbaum 1972) then some future anthropologist would have an abundance of rich materials from American society in 2011. That year, on the cover *Time*, of one of the nation's longest published popular national news magazines and barometer of cultural trends, were two cover stories of interest here: "2045: The Year Man Becomes Immortal" and "What If There's No Hell?" These stories, as will be developed, are part of at least two emerging death systems in the West. Their forms vary owing to differing resolutions to the cultural contradictions between: the embracement of scientism and the resurgence of religious fundamentalism; social successes in generally confining death to the old and the huge public expenditures that are invested in those at life's end (in the United States, one-third spent on healthcare is for those in the last year of life (Gallup International Institute 1997)); science's eschatology of entropy and extinction and yet its promises of a postmortal society; and weakening confidence in an afterlife and yet the resurgence of the dead in popular culture and everyday life.

The Late Modern Ways of Dying

While the two-thirds improvement in our life expectancies would seem to be the occasion for celebration, the nearly-guaranteed new stage of the life-cycle, old age, is increasingly understood to be a "social problem." Not only does this age group carry the stigma of being death-prone, but also its new way of dying challenges traditional social coping mechanisms. The disengagement process is lengthening; older persons now die a number of mini social deaths before actually physiologically expiring. Increasingly their dying occurs in slow-motion due to chronic degenerative diseases, often exhausting the resources and patience of families and society alike. An estimated one-half of those 85 and older, the fastest growing age group in late modern societies, suffer from Alzheimer's disease. *The liminal realm between life and death is shifting from being postmortem to premortem.*

Consider the end-of-life stories of two of the late twentieth-century's most powerful political figures, Ronald Reagan and Margaret Thatcher. The former was diagnosed as having Alzheimer's in 1994 at the age of 83. In a handwritten letter to the nation, he delivered a eulogy for his former self—"I now begin the journey that will lead me into the sunset of my life"—and then disappeared from public view for a decade before finally succumbing to pneumonia. During his slow-motion demise, only a few friends were allowed to visit. Observed his wife Nancy on a televised talk show, "Ronnie would want people to remember him as he was" (CNN 2001). "The Iron Lady," Margaret Thatcher, the victim of dementia induced by a series of small strokes, was at age 76 finally told by her physicians to cease her public speaking. As of the time of this writing, she has occupied the liminal world of her friend Reagan for over a decade.

And when the end of existence finally does occur, rarely is it "good" or "appropriate" but rather a difficult and messy affair (Nuland 1994). More than three-quarters of those at the end of life are unable to make their own medical decisions (Landro 2011). And even if final wishes were recorded earlier, it does not guarantee they will be honored. A 2004 survey revealed that 65 percent of physicians would not necessarily follow a living will if, for example, its instructions conflicted with the doctor's own ideas of the patient's prognosis or expected quality of life (Hardin and Yusufaly 2004).

Dying has shifted from being a moral to a technological rite-of-passage (Kearl 1993). Those in charge are, ironically, not trained to assist in the process. When over 600 patients with advanced cancer and having about six months left to live were recently asked whether their physicians had discussed their wishes for end-of-life care, nearly seven out of 10 said that such conversations had never taken place (Zhang et al. 2009). As a result, according to one major study, one-quarter of dying patients did not get enough pain medication—sometimes none at all (Teno et al. 2004). Another, studying dementia inpatients in a geriatric department of a large American hospital, found that 70 percent of the dying elderly admissions did not die calmly. Seventy per cent had pressure sores and 94 percent were malnourished; 15 percent were screaming and 90 percent had endured an invasive medical procedure one week before their death (Aminoff and Adunsky 2004).

We die when physicians let us die. Two decades ago it was estimated by the American Hospital Association that seven out of 10 deaths in the United States were somehow timed or negotiated. Death now occurs within man-made times, not God's time.

The Movement towards Customized Death

It was because of such developments that individuals began movements to enable one to die one's own death, to return control to the dying and to their families. According to the NORC General Social Surveys (Davis and Smith 2007), when asked if one "has the right to end his own life if this person has an incurable disease?" the percentage of Americans agreeing increased from 39 percent in 1977 to 60 percent in 2008. Over the same time period, more than two-thirds endorsed the question: "When a person has a disease that cannot be cured, do you think doctors should be allowed by law to end the patient's life by some painless means if the patient and his family request it?" Nowadays, those most likely to die dehumanized, McDonaldized hospital deaths are minorities and uninsured patients.

Challenging the technological tortures too often associated with medicalized death are various right-to-die and death-with-dignity movements, whose agendas often include honoring advance directives and DNR requests, and physician-assisted death. The hospice movement, based on a philosophy of palliative treatment and restoration of control to the terminally ill, has gained traction. Further, we note the appearance of the "death tourism" industry.

In sum, the death fears experienced by those most likely to die are shifting from concerns over postmortem judgment to fears of death from intensive care and the well-being of loved ones (Gallup International Institute 1997). A major concern of the old is not so much being dead but rather becoming zombies, the living dead, existing in some persistent vegetative state, totally dependent on others, and being nothing more than a burden on the living. This is not how they wish to be remembered.

The Persistent Faith in an Afterlife

In a 2000 lecture at Darwin College, J.R. Lucas declared: "Religion does not take kindly to time. For many people the religious impulse is generated by a desire to escape from the tyranny of time: we seek a refuge from the changes and chances of this fleeting world, and hope to evade the inevitable mortality of our temporal existence, by moving to a higher plane of reality, where the limitations of temporality are transcended by the eternal verities of absolute existence" (Lucas 2002).

Religion counters the profane times that Lucas was referring to with various sacred times that tap the mysterious, extraordinary and uncontrollable forces that act against chaos (Douglas 1966). These include the notions of eternity and deathlessness, which exist throughout the recorded history of the temporal primate. Religious belief and ritual can thus be understood as largely entailing time-orienting mechanisms (Brandon 1951), breaking up time's continuous flow that spirals into death and extinction and giving followers a supra-biographical orientation to their existence.

The temporal primate is not the only creature to be aware of its own demise but also to feel and express a sense of connectedness with generations of predecessors and successors. Such perspective comprises part of *the sacred*, an awareness of how communal ties are cross-stitched through time in an integrative pattern comprising the cultural fabric. Our separate endeavors, as well as those of our ancestors and successors, are but unrelated discrete events, much like the separate notes of music, unless there's some overarching frame of reference to integrate them into some cohesive whole. Part of religion's role is to give this broader temporal perspective to everyday life, to provide the sense of coherency and comprehensibility to both personal and social experiences, and to place generations now alive in context of those dead and those yet to be born. Arguing how society itself is built upon the death of its members, Joseph Bottum notes how death provides the prototypical reason for association: "It keeps us from drifting by tying us to a temporal reality larger – richer and more significant – than our individual present" (Bottum 2007: 20).

For at least the past two millennia in the West, life's goal has been the achieving of salvation, of being purged of one's sins to enable entry into the eternal paradise of heaven and therein experience awesome joy. As Darryl Reanney thus rightly observed: "By 'inventing' Gods and investing them with the power to preserve life indefinitely, man restored to himself the necessary ability to act purposefully in the face of death" (Reanney 1991: 8). This religious legacy remains profound. As illustrated in Figure 3.1, at least according to this sampling of countries, a majority of humanity is confident in the existence of a life after death. United States is an extreme case among the developed West. There nearly three out of four live with the expectation—greater than all but four of the 35 countries surveyed in 2008 (ISSP 2008). This proportion of Americans has been on the increase at least over the last three decades of the twentieth century, particularly among Catholics, Jews and those with no religious affiliation (Greeley and Hout 1999)—a proportion unexplained by demographic changes (e.g., the aging of the population). When asked in 2010 if they believe that Jesus Christ will return to earth in the next 40 years, 26 percent of Americans believed that it will definitely happen and another 21 percent believed it probably would (Pew/Smithsonian 2010). Among other countries, however, such certainties are waning. When compared to a similar survey conducted in 1998, one finds, for instance, those believing the prospects of an afterlife to be definitely or probably true slipped from 52 to 41 percent in France, 51 to 43 percent in Sweden, 59 to 51 percent in Great Britain, 52 to 46 percent in Norway, and 41 to 37 percent in Denmark.

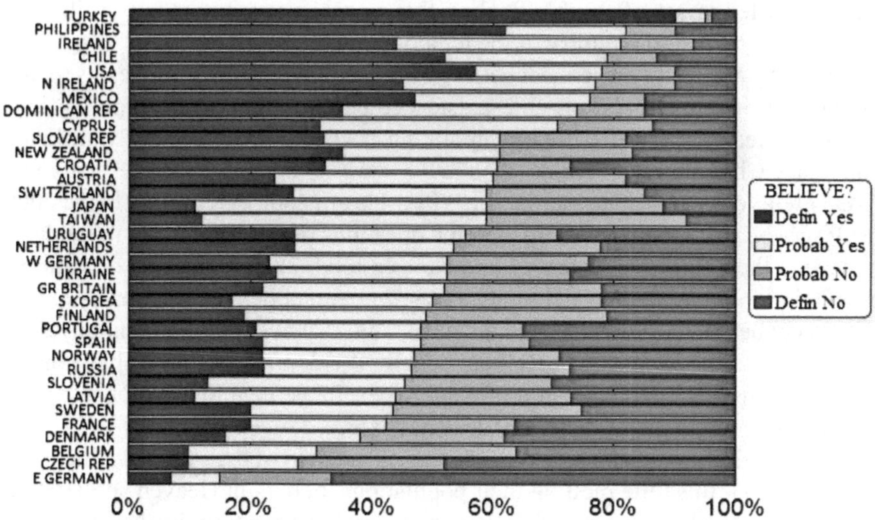

Figure 3.1 Afterlife beliefs across nations

Source: based on data drawn from ISSP 2008.

Secularization, the Fading Allure of the Traditional Afterlife and the Demise of Timelessness

With secularization, dramatic increases in the average life expectancy and the institutionalization of the dying and the dead have fundamentally altered religion's salvific ethos. The goal of living Methuselah-length existences has dimmed in the wake of Alzheimer's disease and other chronic maladies in old age. Within capitalist-based social orders—wherein the core values are youth, speed, productivity and self-sufficiency—the social status of older persons has become problematic. Further, the allure (and social control potency) of traditional transcendent life goals is fading; existence in the here-and-now is what matters most in the highly materialistic West. Radical Islam's promise of 72 virgins to its martyrs seems quaint—and hardly a reward for half the population within its patriarchal societies. Even the eternal attractions of Christianity's heaven are fading in appeal to a populace now conditioned by consumerism and its instant gratifications and continuous stream of novelties. With the marketplace rather than religion now deconstructing death fears, as will be argued, the central existential concerns of highly individualistic, long-lived selves is that of incompleteness, of not having lived full, completed lives.

Secularism has not meant abandoning traditional hopes for deathlessness. Rather, faith in salvation and in eternal life is no longer monopolized by religious institutions. The promises that science will eventually conquer mortality through new death-transcending technologies are believed by many in late-modern secular societies and are increasingly seen to be definitely more verifiable and probable than any similar promise from religion. But admittedly, secularism also can mean the acknowledgment of death's finality, that eternal existence is but an illusion harbored in the neural circuitry of the temporal ape.

We note a growing disconnect between the moral worthiness of lives lived and the quality of individuals' afterlife existence. The traditional monopoly various faiths claimed over the afterlife is evaporating. Two-thirds of religiously affiliated American Christians, according to a 2008 survey, agreed with the statement "Many religions can lead to eternal life" (Pew Forum 2008). Returning to the 2008 ISSP survey, we consider responses to the questions of whether or not individuals believe in the existence of Heaven and Hell. In total, 28 percent of the countries' populations believe Heaven definitely exists, 24 percent believe it probably exists, 18 percent believe it probably does not exist, and 21 percent that it definitely does not. Beliefs in the existence of Hell are equally divided: 24 percent believe that it definitely exists, 22 percent that it probably does, 22 percent that it probably does not, and 31 percent that it definitely does not exist.

In Figure 3.2, we again lay out the countries in terms of their degree of belief in an afterlife but this time plotting their populations' beliefs in Heaven and Hell. In general, with declines in afterlife beliefs in general, we see first the disappearance of Hell and then Heaven as well—Heaven, which previously was the incarnation and materialization of many afterlife beliefs, nowadays increasingly give way to

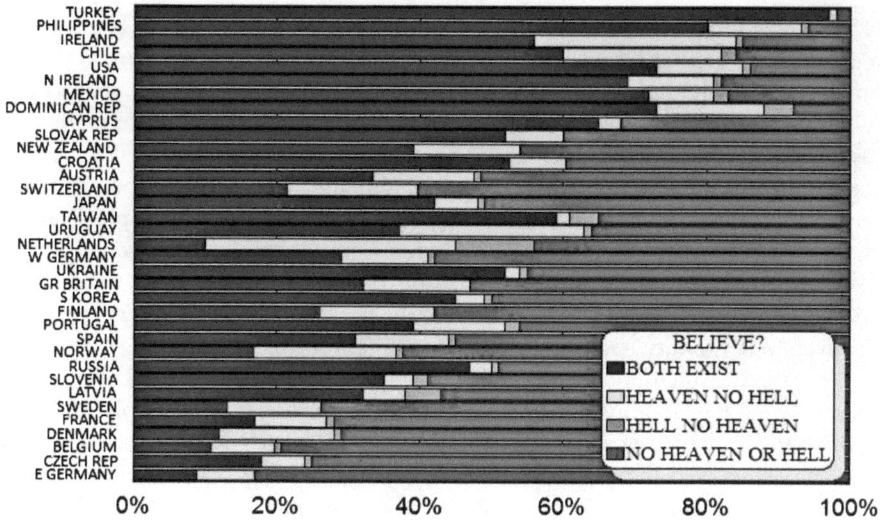

Figure 3.2 Beliefs in Heaven and Hell

Source: based on data drawn from ISSP 2008.

other ethereal and spiritual dimensions (McDannell and Lang 1988). Considered together, 78 percent of those definitely assured about life after death believe in both Heaven and Hell. Of those believing it is at least probable, 59 percent believe in the existence of both afterlife realms and 21 percent believe neither exist. And among those absolutely certain there is no life after death, 81 percent believe neither in Heaven nor Hell. The phenomenon is called "spiritual apathy," shrugging off God, waning concerns over Heaven and Hell, and absence of interest in the meaning of life. Forty-four percent of Americans told Baylor University Religion Survey researchers in 2011 that they spend no time seeking "eternal wisdom," with 46 percent never wondering about whether they would enter Heaven, and 28 percent agreeing "it's not a major priority in my life to find my deeper purpose" (Baylor University 2011).

The Times of the Sciences: Entropy, Extinction and the Dissolution of Eternity

One of the West's cultural shocks of the nineteenth century, on par with that blasphemous insight of Polish priest Copernicus that the earth was not the center of the cosmos, involved the temporal perspective of geologic time. In an era that most thought that the planet was less than 6,000 years old and that it was Noah's flood that carved its geography (leaving, for instance, all of those large round boulders in the more northern latitudes), along came Charles Lyell, who in *Principles of Geology*

(1833) gave evidence that the earth has existed for billions of years. Humanity's place in his timeline became minuscule. In this new chronology there was less opportunity for humans to walk with dinosaurs. In the West, Jesus's appearance two-thirds of the way into the old storyboard of existence now became but an instant before the present. Such are the mental exercises in time that the natural sciences impose. They have also given us the ultimate of temporal perspectives: the birth and death of the entire universe. Death is nothing more than entropic disorder and life nothing more than entropy reduction. Time is the loss of disequilibrium between processes (such as the dissipation of electrical energy or radioactive decay). Human history means very little in the cosmic scheme of things; a human life even less. Eternity became telescoped to unfathomable lengths.

For many contemporary evangelicals and believers in the literalness of the Bible, the understanding is that the world is roughly 10,000 years old and that several Biblical figures lived superannuated existences. Man is created in the image of God, and therefore any morphing of His image through evolutionary processes is unthinkable. Thus, in the strongly religious United States, opposition to Charles Darwin's thesis of evolution remains a cornerstone of the nation's so-called culture wars. Only 26 percent of Americans believe that humans and other living things evolved through natural selection, while 21 percent understand that humans and other living things had evolved over time but under the guidance of a supreme being. More than four in 10 (42 percent) said that humans and other living things have always existed in their present form only (Pew Forum 2008).

Faith in Science's War Against Death

In our late modern societies, humanity's temporal allocations of life owe not to personal sins or ancestral misdeeds but rather are understood to be due to genes, lifestyle choices, and unhealthy social environment. With its medicalization, death has to a large degree been turned into a matter of disease. The U.S. Centers for Disease Control, for instance, describes suicides and homicides, along with influenza, as matters of public health, and monitors their occasional "epidemics." Once identified and labeled, according to the contemporary mindset, diseases will eventually be cured. In highly publicized stories of predicted medical cures, the sciences fan hopes for a forthcoming "no death"-society. Consider the following syndicated headlines collected over the past decade regarding advances against the disease of cancer: "Researchers: New Vaccine Can Stop Lung Cancer," "Dramatic Breakthroughs in Cancer Treatment," "Canadian Team Discovers Gene that Turns Cancers Off" and "Disease Cause is Pinpointed with Genome." Raymond Kuzweil, the reputed Thomas Edison of our times, predicted immortality to be but a few decades away for a software-based humanity. He envisions cybernetically enhanced humans genetically improved through Internet downloads and through whose bodies course billions of "nanobots," molecular-level robots, that reverse the aging process by continuously modifying brain cells, muscles, arteries and bone (Kurzweil 2005). In 2011, British biomedical gerontologist Aubrey de Grey announced that the first person who will live to see her 150th birthday was already alive and that 20 years later, in 2031, will appear one who will live for 1,000 years.

To demonstrate our power against the times of nature, humanity is dedicated itself to preserve endangered species. And, given recent reports of attempts to clone the woolly mammoth, we even try to demonstrate our God-like powers to resurrect the extinct.

In the aforementioned 2010 study wherein a national sample of American adults (n=1,546) were asked about what life would be like in 2050 (Pew/Smithsonian 2010), among the questions asked were whether the following definitely will happen, will probably happen, will probably not happen, or will definitely not happen:

- scientists will bring back an extinct animal species by cloning it;
- we will clone human beings;
- there will be artificial arms and legs that perform better than natural ones;
- there will be a cure for cancer.

	18–29		**30–49**		**50–64**		**65+**		**Total**
	Male	Female	Male	Female	Male	Female	Male	Female	
Clone humans	55	56	54	51	46	46	48	37	49
Resurrect extinct	48	41	49	64	55	53	38	54	52
Artificial bodies	73	66	63	70	64	76	66	60	67
Cancer cure	76	66	68	77	69	69	79	79	72

Figure 3.3 Percentage believing it definitely or probably will happen—by age and gender

For those who cannot wait, whose time has run out before such advancements occur, there exist a host of consolations in various forms of symbolic immortality (Kearl 2010). There are for most, for instance, digitally preserved movies and photographs to document one's existence. To preserve individuals' physical essence have emerged such companies as Geneternity, which stores DNA samples of the deceased, and Alcor Life Extension Foundation, which cryonically suspends bodies at the moment of death. Companies also engage in the business of preserving individuals' social essence, such as Cards From Beyond, which enable the deceased the ability to send cards to loved ones on holidays and anniversaries, and Loving Pup Productions, which provides a posthumous email service.

Accelerating the Dissolution of Corporeal Selves

According to thanatological determinists, at the core of every cultural death system is the manner by which the dead are disposed. Change this ritual and one changes the entire death ethos—which, in turn, is thoroughly interwoven with the broader cultural temporal matrix. In agrarian times, for instance, the burying of the dead just as farmers plant seeds is infused with connotations of resurrection and renewal. As new ideologies emerged to overlay the practice arose Christianity's resurrection of the dead during the Second Coming.

With late modernity we are witnessing new mortuary practices that have and will entail new cultural stances toward death transcendence and time. There has, for instance, across much of the West arisen increasing rates of cremation, which replaces the natural times of decomposition with the clock time of time-accelerating technology. Such disposals of the dead symbolically dramatize late modern Western obsessions to defy nature and to encourage the bereaved to quickly get on with their lives. In 2010, according to the Cremation Association of North America (cited in National Funeral Directors Association 2012), 40 percent of all deaths were cremated in the United States, up from 4.6 percent in 1970. Rates in Great Britain, Denmark and Sweden now exceed 70 percent. In the latter, the search for alternative methods for quick disposal, for ways totally divorced from any associations with the Nazi ovens or Hindi funeral pyres, has led to the development of promessions, where the body is first cooled before being dipped into liquid nitrogen, with the porous remains then pulverized into a biodegradable powder. Glasgow-based Resomation Ltd. offers the service of dissolving bodies in heated alkaline water.

The Contemporary *Ars Moriendi*

Every civilization has created death consolations, which exist as much for survivors as for those facing their ends. Indeed, giving meaning to the meaningless of mortality may be one of the foremost responsibilities of all cultural systems. But unlike the past when mortuary rituals had to address premature death, the present cultural challenge involves post-mature death. Increasingly, individuals die slow-motion deaths due to chronic disease, epitomized by Alzheimer's disease. Lengthy bouts of self-degradation and loss of dignity threaten survivors' memories of the deceased during their prime. In addition, in the context of scientism, notions of eternity have become obsolete. The traditional afterlife consolations decreasingly suffice.

What kind of consolation is needed in a culture where the only certainty is change and where most cannot recall the name of any great-grandparent? While collecting oral histories at retirement communities, one often heard by an author's students is: "I'm glad that I lived when I did. With the way the future looks, I feel sorry for you young people." This consolation, that one has lived during the best of times, is one of several appearing during our era of shortened time horizons.

Customized Opportunities for Death Transcendence: Capitalism Comes to the Rescue

The quest for immortality runs deep throughout the earliest known Western stories (Hentsch 2004), whose themes involve not speculation of the existence of an afterlife but rather how to keep the dead in their place (Walter 1996: 13). With late modernity, the agenda has been reversed, to mobilize the dead to reaffirm opportunities for life after death.

The realm of the dead—e.g. the world of memory—is no longer what it used to be. They no longer reside only within the memories of their contemporaries. With visual and auditory recordings (e.g. photographs, tape and movie recordings, the web), those of the past can still be seen and heard, even by those whose lives never intersected theirs. Increasingly they leave narratives intended for posthumous consumption with the goal of receiving a place on the timeline of social enterprises, an acknowledgment of their part in the histories of nation-states, communities, organizations and families, each of which has become a collective, time- and death-transcending enterprise.

The memories of the deceased are also increasingly attended to by the living. Over the final quarter of the last century, the United States witnessed a veritable explosion in the number of halls of fame as social groups attempt to affirm their sovereign status (hence the national-, state-, city-, organizational- and professional-levels of remembrance) by conferring immortality to their elect. Of the over 200 halls identified from the web, personal contacts, and from Victor J. Danilov's (1997) detailed inventory, all but four were founded since World War II and roughly one-half since 1980. Since the beginning of the twenty-first century, we have witnessed the annual rankings of the earnings of deceased celebrities by *Forbes* and of their recognizability and likeability by Market Evaluations's "Dead Q" listings.

On Bucket Lists and Funerary Shifts toward Life Celebrations

As mentioned, the grief of survivors traditionally involved the frustrations surrounding premature deaths—evident in such traditional consolations as "the good die young." Collective senses of the injustices of lives so curtailed have become amplified with their relative scarcity. Nowadays, among the most successful of charities are those of the "make a wish" movement, where a major desire of a terminally ill child—such as visiting Disneyland or meeting a sports celebrity—is granted. Similar charities do not exist for those at the other end of the lifespan. Nor, increasingly, do their funerals resemble each other.

According to industry informants, as death predominately comes to those having lived full, completed lives, funerals are morphing from standardized rituals of collective grief to customized celebrations of lives well lived, from religious narratives on the virtues of an afterlife to therapeutic entertainment. A Harris poll, commissioned by the National Funeral Directors Association, found that 62 percent of the American Boomer generation desire some kind of personalized funeral, from themed caskets to graveside releases of butterflies or balloons (Carpenter 2005).

Also morphing is the century-old "living memory" staging of the funeral industry, featuring the embalmed body within an open coffin. Displayed was the individual at the time of his death, idealized cosmetically to erase end-of-life agonies. But nowadays caskets are increasingly closed. Instead, the memory that is sought to be preserved is increasingly that of the deceased at their prime of life, not the aged self-caricature whose departure from life arrives postmaturely. Longitudinal analyses of newspaper obituaries over the last third of the twentieth century time reveal a strong trend toward "age-inaccurate" images of the deceased, showing not those who died but rather the images of their much younger selves (Anderson and Han 2008–9). In sum, the industry is shifting from preserving corpses to preserving the memories of the deceased at their prime.

During the first decade of the twenty-first century appeared the concept of the "bucket list," detailing the experiences and accomplishments one wishes to have before one dies. The origin of the phrase comes from the title of a 2007 movie (with Morgan Freeman and Jack Nicholson performing the roles of the terminally ill) featuring two terminally-ill elderly men who travel around the world doing all the things they have ever wanted to do before they perish. Capitalism is more than ready to define and market such wishes. The movie spawned a genre of so-called "… Before You Die" books, with the blank filled in with such notions as books you must read, movies one must see, "must" places to visit, wines you must taste, places to play golf, etc.

Such phenomena, we contend, are bound up with a new death ethos, one perfectly in tune with the course of late modern economies. Here the ideology holds that death comes to those who have lived fully and who have completed their living. The compensations of some eternal afterlife are no longer necessary. And, just as we are more likely to remember stories that are unresolved, so are we ready to let go and to forget those whose biographies are complete. With fully checked-off bucket lists, we have a tangible measure of completeness to share at funerals. The items themselves are largely defined by the market.

Conclusion

At least since the Middle Ages, and compared to other so-called "more primitive" cultures around the world, Western Man has always struggled to find meaning in death (Choron 1973; Carse 1980). As Harry Redner once provocatively observed on the irony and paradox of mankind's increasing desire to abolish death: "Man is now mortal, precisely by having acted on the premise of Man's immortality" (Redner 1982: 13). In this chapter, we have attempted to illustrate and discuss some of the ironies, paradoxes and absurdities involved in our contemporary late modern "death system's" adamant aspirations and determination to deal with—and if possible "kill off"—death and to control the span of life being lived prior to death. Instead of physiological death and material decay being the key temporal referents of social life, in late modern societies the temporal referents are now becoming social death and the shortening half-lives of memories of individuals' existences. With weakening faith in the existence of some sacred eternal realm beyond death and with the proliferation of "premature" and particularly "postmature" deaths, the liminalities between life and death are shifting from the afterlife to late life. As the essence of selfhood is increasingly symbolic, concerns over preserving an embalmed corpse for resurrection are redirected toward the immediate destruction, through cremation, of the diminished self that threatens to contaminate memories of the deceased in his or her prime. And with the new technologies that allow the postselves to be seen and heard by others far removed in time, the illusions of death transcendence and permanence are preserved.

References

Aminoff, Bechor Zvi and Abraham Adunsky (2004) "Dying Dementia Patients: Too Much Suffering, Too Little Palliation." *American Journal of Alzheimer's Disease*, 19(4): 243–7.

Anderson, Keith A. and Jina Han (2008–9) "An Exploration of Ageism and Sexism in Obituary Photographs: 1967–1997." *Omega: Journal of Death and Dying*, 58(4): 335–45.

Bauman, Zygmunt (1992a) *Morality, Immortality and Other Life Strategies*. Stanford: Stanford University Press.

Bauman, Zygmunt (1992b) "Survival as a Social Construct." *Theory, Culture & Society*, 9: 1–36.

Baylor University (2011) "The Values and Beliefs of the American Public: Wave III Baylor Religion Survey." Retrieved from http://www.baylor.edu/content/services/document.php/153501.pdf (September).

Becker, Ernest (1973) *The Denial of Death*. New York: Free Press.

Becker, Ernest (1975) *Escape from Evil*. New York: Free Press.

Bottum, Joseph (2007) "Death & Politics." *First Things*, 174 (June/July): 17–29.

Brandon, S.G.F. (1951) *Time and Mankind*. London: Hutchinson.

Carpenter, Mackenzie (2005) "Baby Boomers find Solace in Customized, Multimedia Funerals." *Pittsburgh Post-Gazette*, January 14.
Carse, James P. (1980) *Death and Existence: A Conceptual History of Human Mortality*. New York: John Wiley and Sons.
Choron, Jacques (1973) *Death and Western Thought*. New York: Macmillan.
CNN.com transcripts (2001) "Nancy Reagan Reflects on Life With Ronald" (March 4). Retrieved from http://transcripts.cnn.com/TRANSCRIPTS/0103/04/lklw.00.html.
Danilov, Victor J. (1997) *Hall of Fame Museums: A Reference Guide*. Westport: Greenwood Press.
Davis, James Allan and Tom W. Smith (2007) *General Social Surveys, 1972–2006*. National Science Foundation, Chicago: National Opinion Research Center.
Douglas, Mary (1966) *Purity and Danger*. New York: Praeger.
Elias, Norbert (1985) *The Loneliness of the Dying*. London: Continuum.
Falk, Dan (2008) *In Search of Time: The Science of a Curious Dimension*. New York: Thomas Dunne Books/St. Martin's Press.
Gallup International Institute (1997) *Spiritual Beliefs and the Dying Process: A Report on a National Survey* (conducted for the Nathan Cummings Foundation and the Fetzer Institute). Princeton: The George H. Gallup International Institute.
Goffman, Erving (1967) *Interaction Ritual*. New York: Pantheon Books.
Greeley, Andrew and Michael Hout (1999) "Americans' Increasing Belief in Life After Death: Religious Competition and Acculturation." *American Sociological Review*, 64: 813–35.
Hardin, Steven B. and Yasmin A. Yusufaly (2004) "Difficult End-of-Life Treatment Decisions: Do Other Factors Trump Advance Directives?" *Archives of Internal Medicine*, 164: 1531–3.
Hentsch, Thierry (2004) *Truth or Death: The Quest for Immortality in the Western Narrative Tradition*. Vancouver: Talonbooks.
Hick, John (1976) *Death and Eternal Life*. San Francisco: Harper & Row.
International Social Survey Program (ISSP) (2008) *International Social Survey Program: Religion III*. Cologne: Zentralarchiv für empirische Sozialforschung.
Jacobsen, Michael Hviid (2012) "The Cultural Construction and Deconstruction of Death," in Michael Hviid Jacobsen (ed.) *Deconstructing Death – Changing Cultures of Death, Dying, Bereavement and Care in the Nordic Countries*. Odense: University Press of Southern Denmark.
Kastenbaum, Robert (1972) "On the Future of Death: Some Images and Opinions." *Omega: Journal of Death and Dying*, 3: 306–18.
Kearl, Michael C. (1993) "Dying American Style: From Moral to Technological Rite of Passage." *American Journal of Ethics and Medicine*, 2(1): 12–18.
Kearl, Michael C. (2010) "The Proliferation of Postselves in American Civic and Popular Cultures." *Mortality*, 15(1): 47–63.
Kurzweil, Raymond (2005) *The Singularity is Near: When Humans Transcend Biology*. New York: Viking.

Landro, Laura (2011) "New Efforts to Simplify End-of-Life Care Wishes." *Wall Street Journal*, March 15.

Lucas, J.R. (2002) "Time and Religion," in Katinka Ridderbos (ed.) *Time*. Cambridge: Cambridge University Press, 143–68.

McDannell, Colleen and Bernhard Lang (1988) *Heaven: A History*. Yale: Nota Bene.

National Funeral Directors Association (2012) "Statistics." Retrieved from http://www.nfda.org/media-center/statisticsreports.html#cfacts.

Nuland, Sherwin (1994) *How We Die: Reflections of Life's Final Chapter*. New York: Knopf.

Orme, John Edward (1962/9) *Time, Experience and Behaviour*. London: Iliffe Books Ltd.

Perrett, Roy W. (1986) *Death and Immortality*. Dordrecht: Kluwer Academic.

Pew Forum on Religion and Public Life (2008) "Many Americans Say Other Faiths Can Lead to Eternal Life." Retrieved from http://pewresearch.org/pubs/1062/many-americans-say-other-faiths-can-lead-to-eternal-life.

Pew Research Center/Smithsonian Institute (2010) "Life in 2050: Amazing Science, Familiar Perils." Study retrieved from http://www.people-press.org/files/legacy-pdf/625.pdf.

Phillips, David (1972) "Deathday and Birthday: An Unexpected Connection," in Judith M. Tanur (ed.) *Statistics: Guide to the Unknown*. San Francisco: HoldenDay, 52–65.

Phillips, David and Kenneth A. Feldman (1973) "A Dip in Deaths Before Ceremonial Occasions: Some New Relationships Between Social Integration and Mortality." *American Sociological Review*, 38: 678–96.

Prioreschi, Plinio (1990) *A History of Human Responses to Death: Mythologies, Rituals and Ethics*. Lewiston: Edwin Mellen Press.

Reanney, Darryl (1991) *The Death of Forever – A New Future for Human Consciousness*. London: Souvenir Press.

Redner, Harry (1982) *In the Beginning was the Deed*. Berkeley: University of California Press.

Teno, Joan M., Brian R. Clarridge, Virginia Casey, Lisa C. Welch, Terrie Wetle, Renee Shield and Vincent Mor (2004) "Family Perspectives on End-of-Life Care at the Last Place of Care." *Journal of the American Medical Association*, 291: 88–93.

Toynbee, Arnold (1980) "Various Ways in Which Human Beings Have Sought to Reconcile Themselves to the Fact of Death," in Edwin Shneidman (ed.) *Death: Current Perspectives*. Palo Alto: Mayfield, 11–34.

Walter, Tony (1996) *The Eclipse of Eternity*. London: Routledge.

Zhang, Bahui, Alexi A. Wright, Haiden A. Huskamp, Matthew E. Nilsson, Matthew L. Maciejewski, Craig C. Earle, Susan D. Block, Paul K. Maciejewski and Holly G. Prigerson (2009) "Health Care Costs in the Last Week of Life." *Archives of Internal Medicine*, 169(5): 480–488.

Chapter 4
Rebirth and the Death Drive: Rethinking Freud's "Mourning and Melancholia" through a Siberian Time Perspective

Rane Willerslev

This chapter is concerned with so-called rebirth beliefs in the Siberian North—that is, the cosmological principle that a dead person's soul will circle back to be reborn in a neonatal, who can, therefore, be identified as an ancestor returned. This effectively means that all living people are not simply themselves, but also always some returned deceased ancestors as well. What does this principle of rebirth with its cyclic understanding of time offer its practitioners? For we may ask: does the embodied presence of the dead signify an altogether different psychological make-up than the one commonly known to Western psychology? One possible answer, which I will pursue here, is that the Siberian rebirth doctrine offers a particular model for dealing with grief, which emphasizes the continuity rather than finality of personal relationships. I will show this by comparing the Siberian indigenous model with Sigmund Freud's (2009 [1917]) famous study of how a bereaved person may, in the face of loss, regress into a depressive state of melancholia, something which Freud considered a severe and even pathological response to death that can result in a suicidal identification with the deceased. I will reveal how the Siberian rebirth beliefs offer a quite different type of relief to the pain of loss; a difference which boils down to a distinct conception of time and its impact on what counts as reality. My aim with the comparison is not merely to criticize Freud by revealing how his theory fails to include so-called non-Western psychologies. Rather, my aim is to use my Siberian ethnography to unlock the secret treasures of Freud; to help taking his psychoanalytic theory to its full analytical potential.

I am especially concerned with contributing to a better understanding of the meaning of Freud's perhaps most ill-understood concept of *Thanatos*, the so-called "death drive." By revisiting the death drive through the outlook of the Siberian rebirth eschatology and coupling it with the philosopher Gilles Deleuze's theory of time, I shall end up rereading the death drive in a manner that is precisely contrary to most prevailing interpretations; namely as an indestructible insistence of the libido—an excess of life that persists beyond the biological cycle of life and death. If my capacity to arrive at this reinterpretation of the death drive is satisfactory, then it suggests that the two models—the indigenous and the psychoanalytic—may indeed be considered as mutually enlightening discourses that can provide important insights into each other. This kind of comparative dialogue has only been little exploited to date (Vitebsky 1993: 16) and it can point to a way to effect a more fruitful cooperation between anthropology and psychoanalysis, where they are not seen as simply opposed or even complementary visions, but creatively feed back into each other's disciplinary perspectives.

Siberian Rebirth

In a collection of essays named *Amerindian Rebirth* (Mills and Slobodin 1994), which describes rebirth beliefs among the American Indians and the Inuit, a number of scholars reveal convincing evidence that such beliefs were prevalent throughout virtually all of North America. Moreover, the anthropologist Gananath Obeyesekere (2002: 37) has suggested that "these ideas were probably part of an ancient, wider circumpolar cultural complex … stretching into all of central Asia." Still, with the exception of a few sketchy studies (see Bogoras 1904–9; Jochelson 1926; Vitebsky 2005; Willerslev 2007), the north Asian region, or what is more popularly known as Siberia, has remained practically undocumented in relation to questions of rebirth beliefs and it needs the kind of deeper exploration conducted by the scholars of the Northwest Coast and Inuit religions.

In this chapter, I take up the challenge of documenting and interpreting the rebirth beliefs among two northeastern aboriginal populations, the Yukaghir and

Figure 4.1 Elderly Yukaghir woman, sewing in a tent

Figure 4.2 Chukchi couple in front of their tent

Chukchi, with whom I have done long-term fieldwork (Willerslev 2007, 2009, 2012). Besides speaking distinct languages that belong to different language groups, the greatest difference between the two groups rests in their modes of subsistence and environment of residence: whereas the Yukaghir inhabit the forest region around the Kolyma River and live almost exclusively by hunting and fishing, the Chukchi inhabit the treeless tundra of northern Kamchatka and Chukotka and live either as large-scale reindeer herders in the interior or as hunters of sea mammals on the Arctic seashore.

Despite these obvious differences, the two groups share many common aspects of cosmology, especially with regard to their rebirth doctrines.[1] Both thus subscribe to the notion that all living people are returned deceased relatives that have already passed through several lifetimes. Since the living are considered embodiments of

1 It is important to notice that the forced adoption of Christianity by many Siberian natives in the eighteenth century (not the Chukchi, however, who have largely remained pagans) and the impact of 70 years of Communism have complicated the form and features of the Siberian rebirth eschatologies. Today, not everybody ascribes equal weight to this kind of spiritual knowledge. However, as testified by the vivid accounts I recorded, the power of these beliefs, even in the face of competing ideological and religious principles, are still a very real and genuine dimension of many peoples' daily existence.

the dead and their former lives are understood quite literally to pass through the past lives of their ancestors, there is an important sense in which time among the Yukaghir and Chukchi is understood in terms of cycles of endless repetition.

People are, so to say, always already in the past. The past, then, is not something that comes after the present has ceased to be; rather, the past is always impregnated in the present, even as the present becomes past. The difficult thing to understand, however—and which I will discuss in greater detail later—is that this repetitious time does not consist in the return of the same, but of the different, a kind of "return of difference."

Anthropology and Psychoanalysis

To obviate any possible misunderstandings, I need to say a few things about my use of psychoanalysis, before continuing my interpretative endeavor. While early anthropologists, such as Robert Hertz (1960 [1907]), W.H.R. Rivers (2001 [1932]), C.G. Seligman (1924), Bronislaw Malinowski (1929) and, above all, George Devereux (1978a, 1978b), had few hesitations in using psychoanalytic ideas to interpret their fieldwork data, the later relationship between anthropology and psychoanalysis has remained ambivalent at best, not to say conflictive (Heald et al. 1994). Anthropology clusters much of its distrust of psychoanalysis around two complaints, its presumed ethnocentrism and reductionism: it is argued, respectively, that psychoanalysis at best reveals the workings of the Western mind (Paul 1980), and that it reduces everything to questions of childhood fantasies (Born 1998). Why then do I wish to bring in a psychoanalytic framework, which for the most part has been developed within the privacy of clinical therapy? Is a psychoanalytically informed interpretation really a productive way of engaging with Siberian understandings of death, time and the person?

First, it is worth noting that what I am proposing is not a return to the kind of crude oversimplification of indigenous concepts by psychoanalytic explanations of which there have been a number of cases (see, e.g., Whiting 1953; Spiro 1982; LeVine 2001) and toward which many anthropologists have remained rightly suspicious (see, e.g., Gellner 1985: 5). The potential benefits of bringing the two disciplines together lies, therefore, not in the expectation of revitalizing the reductionist dream of mapping out the psyche of mankind per se. Rather, it lies in the moment of comparison itself—the reading of one discipline in the light of the other (see Schrempp 1992: 5–6). Such a comparative act contains the potential of upturning our assumptions in both disciplinary directions. If Siberian peoples tell us that persons are recycled in endless rebirths, the scholarly exercise should not be to make a simple one-to-one translation of this idea into pre-established psychoanalytic concepts of "repetitive desire," "wishful psychosis" and so on; by such a move, the creative potential of indigenous metaphysics gets at once "castrated" (to stay with Freudian terminology), posing no challenges to our ontological certainties, and we can get on with our work without having to worry about whether there is any foundation in reality for what people are telling us. Instead, the task must be to question our Euro-rationality about the nature of time, death and the person, so as to make it possible for us to imagine how rebirth actually is a fact of life (see Holbraad 2007: 204; Willerslev 2007: 3), which in turn may help illuminating psychological queries within our own Western community. Only by taking indigenous metaphysics seriously in this way can we begin to discuss the complex pattern of common features and differences between the psychoanalytic models of explanation and the principles of indigenous cosmology.

Having said this, it is important to stress that I take the reader with me to a developing argument, not a finished one. One should, therefore, not read my analysis as a fully concluded study of what a psychoanalytically minded ethnography should or must look like. This is a tentative attempt.

The Recycling of Souls

The word "reincarnation" comes from Latin roots, meaning "again embodiment," which implies that the human's soul lives on from life to life in a new physical body. This idea is also reflected in the Siberian peoples' own ideas about rebirth. Thus, the Yukaghirs say: "Shoromoayibiikel'eilio'nin," which means: "A person's shadow (soul) comes back in this person's body." The Chukchi simply call it "Ra'htIlm", meaning "the one coming home." However, this is virtually the only thing that the Siberian rebirth beliefs have in common with the more well-known religions of reincarnation from Southeast Asia. For the Siberian peoples, the objective of rebirth is not to get out of the circle of reincarnation, as in the Buddhist or Hindu tradition, where one tries to attain "nirvana" or "moksha," the Sanskrit word for salvation. Instead, the Siberian peoples expect that people must pass through infinite rebirths and bring back personality traits that they displayed in their earlier lives. Buddhists and Hindus see reincarnation as dependent on how one behaves in pure moral terms in this world, but I have not found any corresponding idea in Siberia that sinful actions in this life will lead to punishment in the next (Willerslev 2012: 164). A dead person can be transformed into an animal or get trapped in places on the journey, but eventually he will return to the "ordinary" world he left and there are no fixed ideas about good and bad rebirths, just as there are no conceptions of a heaven or a hell. The deceased will have an intermediate sojourn in the afterworld of the ancestors and spend another lifetime there, before subsequent rebirth, but this ancestral society is more or less a continuation of everyday life on earth. Moreover, it is not just animate beings, humans and animals, but also inanimate objects that are confined within their own circle of continual rebirth (Willerslev 2007: 32). As anthropologist Lee Guemple (1994) writes for the Inuit, and which seems equally true in the Siberian North, nothing is excluded from the system and nothing new will be able to enter into it. In principle at least, there is a fixed pool from which souls circulate endlessly from this world to the next and back again.

Naming the newborn is perhaps the single most important event in the Siberian rebirth cosmologies. The infant will be given the name of his or her presumed incarnation and the name is, as anthropologist Barbara Bodenhorn (1989) describes for the Inuit, endowed with spiritual power. In fact, there is an important sense in which the name itself is equivalent, if not identical, with the incarnated soul: while its bearer may live or die, the name will go on forever. If a child's identity is not correctly determined and it is given the wrong name, the incarnated soul will be offended and not like its owner. So it will work against the person rather than helping him in life by making him ill and preventing him from luck in hunting and herding (Willerslev 2007: 60). For this reason, the Siberian peoples are very attentive toward determining their infant's true identity. Now and then, a dying person will disclose in advance the pregnant woman he will attempt to enter. At other times, a woman will dream about a definite deceased person during her pregnancy, and in this way she will know what relative has returned in her unborn child. In other instances, the child will, when it is two or three years old and has just learned to speak, declare who it is by describing events from its previous life.

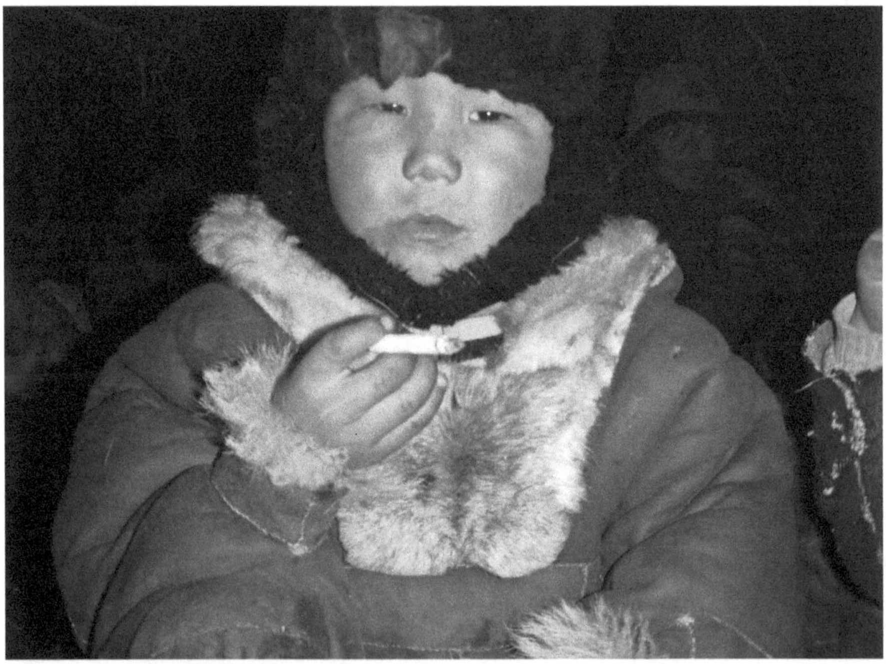

Figure 4.3 Chukchi boy

When he was three years old, he asked his grandfather for a cigarette and was given one since the boy is considered to be the incarnation of the latter's brother, who was a chain smoker. Since then the boy has been smoking heavily just like his namesake.

Basically, all of this implies that any given person is never just him or herself, but always a dead ancestor incarnate as well. “The two are one and the same person,” a Yukaghir assured me. For this reason, formal kinship terms and concomitant relationships may be rearranged when the identity of a newborn has been established. It is also for this reason that Siberian peoples generally say that living a life is a long act of memory, remembering what one already knows (Willerslev 2007: 161). Still, matters are not as simple as they may seem at first. We are not simply dealing with a faithful doubling of personhood, but with something far more complex. First, the same ancestor may be reborn simultaneously as different individuals, so that the same deceased person lives on in two or more people from the same community, sometimes even within the same family. Among the Chukchi, I was even told that “any one person can contain up to four dead people within, although one of them will be the dominant one.” So, in this case there is a primary incarnate and several secondary ones constituting the persona. This produces a kind of “dividuality” (Strathern 1988: 13), in that the person is not the embodiment of a deceased in particular, but is constituted out of character traits of a multiplicity of ancestors. Among the Chukchi, it is also common to find cross-gendered rebirths, so that, for example, a male child is an incarnation of a female relative and will, therefore, be given her female name (see Nuttall 1994: 128; Bodenhorn 1989 for similar ideas among the Inuit). Furthermore, some people would add, “a person’s body is him or herself,” thus emphasizing that it is an aspect of the person, the soul, rather than the person *in toto* that is reborn. What all of this quite clearly points to, is that a person is never simply identical to or exactly the same as his incarnation(s). He is also a person in his own right. Thus, “duplication” or “copying” seems to be something of a misnomer: rebirth consists in the return, not of the identical, but of someone *other*. With respect to time, this signifies the curious notion that every new difference is also the moment of an immediate recurrence. In other words, difference and repetition are not seen in opposition: what is produced, the absolute new itself, is in turn nothing but repetition (Deleuze 1994: 90). Indeed, this is also the critical idea in Deleuze’s *Difference and Repetition* (1994), which I shall discuss later. For now, we need to turn to Freud’s thesis on mourning and melancholia.

Freud’s Thesis

In his essay “Mourning and Melancholia” (2009), Freud explored the individual’s reaction to an actual loss of a loved person, or the loss of an ideal. Why, Freud asks, do some people respond with the affect we call “mourning,” which will be overcome after a certain time, while others sink into depression—a syndrome Freud calls “melancholia.” Freud concludes that a favorable progress of mourning—that is, toward its resolution—implies a detachment from the “love object” and an acceptance that acts of love can no longer be returned. Freud denotes it “reality-testing” (2009: 244) by which he means that the “reality principle” prevails, and the bereaved accepts the loss as irreversible. Psychologist, Veikko Tähkä (1984), in a more recent study of grief-relief, reaches effectively the same conclusion, when he speaks of successful mourning as a process in which the mourner turns a "hot" mental representation of the lost object into a "cold" one by making it "futureless." Both accounts infer the emotional acceptance that the lost object is no longer there to respond to the mourner's wishes, has no future and therefore no ongoing perpetuating influence. So, while psychoanalytic thinking accepts that the bereaved may deny for a while that a loss has actually taken place and wish to have it reversed, he must sooner than later give up this "unhealthy phantasy"; killing off the memories of attachment to the lost love object as a means of re-establishing mental health and returning to life (Freud 2009: 244). Only then, Freud writes, "the ego becomes free and uninhibited again" (2009: 245).

At the other end of healthy mourning stands the melancholic person, who remains sunken in his loss. Freud equates it with the compulsion to repeat; an uncanny urge to go on repeating painful memories. For Freud, repetition is thus the inadequate means by which the bereaved seeks to master the agonizing forces of loss, which exceed his control. But for Freud repetition is a self-destructive reaction: instead of accepting the "call of reality," which blocks the otherwise infinite and boundless repetition of sad memories, the melancholic starts "narcissistically identifying" with the lost love object (Freud 2009: 252). Accordingly, the love and hate ambivalence that originally connected the bereaved to the lost person is turned inward as an assimilated mental representation. Freud insists that the bereaved immediately needs to shift his attention away from the memory of the departed and "withdraw his libido," since it is only by distancing himself from any repetitive re-enactments that its power to harm him is diminished. If the melancholic collapses under the weight of his own ambivalent identification with the deceased, he will eventually lose interest in staying alive and commit suicide. "In mourning," Freud writes, "it is the world which has become poor and empty; in melancholia it is the ego itself" (2009: 246).

It is important to note that Freud's study of melancholia and the pull of death that it entails set him on the trail that led him to discover the so-called "death drive"—that is, Freud's later claim that "the aim of all life is to die ... to cancel itself out" (1961: 39). Freud's thesis of the death drive remains the most controversial and contested of his many concepts. Indeed, for the mainstream of the analytical movement, the death drive has been considered a kind of conceptual "madness," lacking fundamental grounding in empirical observation and research, and it has largely remained outside the canon of accepted Freudian terms (Boothby 1991: 7). It holds true that Freud did not give us any clear account of how the death drive is to be understood or where it is placed in relation to the rest of his work. He tried at times to situate it within biology, talking about it as the organism's "impulse toward returning to an inorganic state" (Freud 1961: 32). At other times, he referred to it as the "nirvana principle," the reducing of internal psychic tensions to an absolute minimum (Freud 1955: 56). In any case, the death drive is arguably "the darkest and most stubborn riddle posed by the legacy of psychoanalysis" (Boothby 1991: 1).

It amounts to saying that the true goal of living is dying. Does it simply refer to something as common as our "waking lassitude," the fact that sometimes we "can't be bothered to face the day," as suggested by Josh Cohen (2005: 104)? If so, this would ultimately manifest itself in the urge to kill ourselves at the first available opportunity, which quite frankly is an absurdity. I think that Freud had almost the opposite in mind when thinking of the death drive, something which I believe that the Siberian rebirth cosmologies can help us in illuminating. Freud did not tell us the exact meaning of the death drive and perhaps he was not fully aware, but I shall venture to make a suggestion on his behalf at the end of this chapter.

Two Case Studies of Rebirth

For now, I will turn to Siberia, where we find a rather different understanding of grief-relief, centered on a kind of perpetual mourning, yet without inevitably leading into melancholia. Let me begin by quoting two rebirth accounts which will then discussed in comparison with Freud. The first was recorded among the Yukaghirs and it was told to me by an elderly woman, Irina, and it is about the relationship to her grandson, Pavel (for more details, see Willerslev 2007: 52):

> The winter night my grandson Pavel was born, I had a dream. I saw my [dead] husband standing naked in front of the oven. He was shivering all over and his skin was blue with cold. The very same night, a helicopter arrived from Zyrianka [the regional centre] to take my daughter to the hospital. You see, major complications followed the delivery. She was unconscious when they transported her, so they completely forgot about Pavel [the infant], who was lying almost naked in the helicopter. He nearly froze to death. Already then, I suspected that he [her husband] had come back [as an incarnation in her grandson] … At that time, I worked in the kindergarten. Pavel comes up to me all the time and wants to help me with different things. During the breaks, he also wants to lie beside me in the bed … When he was five years old, he said to me: "Do you remember those boots that you lost while we were hunting in the forest? I found them and put them in a tree together with our food …" Everything he told me was true. I was very upset and went to my daughter and asked if she had been telling him these things. But she said no, she hadn't said anything … Sometimes when I hear him talking, I hear his [her husband's] voice. [She starts to cry.] Today, he [Pavel] is twelve years old. He brings me all the hares and fish that he catches. The way he hunts is also the same … and the way he talks … he [her husband] has come home to provide for me … I dress him like I used to do and mend his clothes.

I recorded a somewhat similar story, concerning the reunion of a grieving Chukchi man and his deceased wife through the latter's return in a little girl, in the village of Achayvayam in northern Kamchatka. It was an elderly woman, Nina, who explained it to me.

> An old man called "Hairy Bullocks"[2] lost his wife and is now alone. He regards my brother's youngest daughter Tanya as his returned wife. The rest of the village also knows this to be the case. Even when Tanya was little, she would walk up to him and invite him home for tea. Hairy Bullocks would sit there and smile as she pulled him a mug. He would respond by saying: "Thank you, my wife", then sniff her under the nape of her neck and under the chin [a sign of affection]. We would all laugh. At some point, it was told that Hairy Bullocks had been beaten by a group of drunken boys. Tanya, who was five at the time, got all upset. She cried as she ran to his home to make sure that the old man was alright. The two have without a doubt a very close relationship. They care deeply for each other.

The anthropologist Mark Nuttall (1994: 128–9) informs us that in traditional Inuit society the newborn child would literally take the position within the kin group of the deceased person of whom he was considered an incarnation, and be addressed by the same kin terms as they would normally have applied to their dead relative. Among the Yukaghir and Chukchi, we do not find such a formal slant to their rebirth doctrines, at least not at the present day. I did not, for example, ever hear Irina address Pavel directly as her "husband." Although I did hear formal kinship terms used for incarnates among the Chukchi—as when Hairy Bullocks calls Tanya his "wife"—it was not so much a prescribed rule of conduct as it was a manner of speaking in particular contexts of intimacy and joking. Still, both stories clearly reveal that, on the emotional level, there is a strong sense in which the original bond between the deceased and the bereaved is re-established. Irina, for example, tells us that her husband "has come home to provide for her." Therefore, while she calls Pavel her "grandson," she also sees him as her returned "husband." This also goes for Hairy Bullocks, when he addresses Tanya as his "wife." It does not imply that he sees the little girl as his wife in any absolute sense. Rather, he adopts an ambiguous view on the girl as both "wife" and "grandchild," something which is emphasized by the humor underlying the situation.

The Double Perspective

This kind of "double perspective," which signifies that the child confronted swims in and out of focus as both a returned deceased relative and an ordinary grandchild, is of paramount importance for coming to grips with the Siberian rebirth model as grief-relief. Although it may appear totally alien to us, its dual principle may actually lie at the heart of the etymological meaning of our word "mourning." Carlos Mario Aslan (2009: 168) has thus pointed out that mourning translates

---

2 The original Chukchi names are of this kind and not flattering. They are the names of the cannibal spirits, *ke'let*, and are meant to trick these spirits into believing that the person so named is a fellow being and not human prey.

into Spanish as "duelo," derived from Latin, and in this latter language it has two etymologies: (a) from "dolus," meaning "pain," and (b) from "duellum," meaning "duel" or "combat" ("due," meaning "two" and "illum," from "bellum," meaning "war"). Whereas we customarily think of pain as integral to mourning, the second variety, the more obscure notion of a "duel" or "combat," has been less explored.

However, the connotation to an inter-psychic "duel" or "combat" between perspectives is exactly what the Siberian narrations point to: the bereaved is forced to hold two mutually exclusive, yet jointly held, perspectives in view which, therefore, come to coexist in a kind of unresolved dialectical tension. In other words, we are dealing with a kind of inner struggle between perspectives, the fragile balance of which is of the greatest importance for the interaction to be successful. None of the perspectives must be allowed to entirely take over and dominate the other, if the bereaved is to fill the void left by the death of his loved one. Were the bereaved, for example, to reduce the child to an absolute replica of the deceased it would, as I will describe below, be regarded as an act of madness by the community and a violation of the child's autonomy. Hence, it is exactly the capacity to steer a difficult course between transcending difference and the necessity of maintaining sameness that is key to the success of rebirth as grief-relief.

If we turn to a comparison with Freud, it is striking that contrary to his claim to successful mourning, the Siberians keep their mental representations of their lost love object "alive" and "hot" (Tähkä 1984). Thus, contrary to the instructions of psychoanalysis, they simply keep investing their "libidinal cathexis," to use Freud's term, in the dead, who are considered alive and permanent. Yet in the two case-stories described, none of the bereaved becomes a captive of the pain of melancholic desire and the pull of death. Instead, they appear to experience a significant degree of contentment, exactly because they do not let their lost love object go.

Two Siberian Case Studies of Melancholia

Does this then mean that a neurotic state akin to Freud's melancholia does not exist among the Siberian indigenous peoples? I venture to say that it does and that they may actually have ideas that correspond to the Freudian distinction between normal and pathological mourning. Let me describe a couple of cases, which are quite comparable to the ones depicted by Freud.

Among the Yukaghirs, I came across a relationship between a grandfather and grandson that had taken a quite tragic turn (for more details, see Willerslev 2007: 53). A man, Sergei, lost both of his sons within a period of a few years. One drowned, and a young relative killed the other while drunk. Sergei now considers his daughter's son, Vasili, to be the reincarnation of one of his dead sons. But people say that he has not got over the death of his sons and that grief has driven him to insanity. "He has lost his senses," his wife explained to me. "He won't leave Vasili alone … The boy is both nervous and afraid of his grandfather."

I attempted to set up an interview with Vasili to ask him about his sense of identity and reactions to being named after his dead relative, but on the day of our meeting he had left the house. His mother apologized to me for her son's disappearance, but said: "Vasili doesn't believe that he is my dead brother. It is my father who has gone mad. He doesn't understand that nothing can bring back his son." Later that year, the depressed Sergei walked into the forest, never to return. People say that the grief made him commit suicide.

What is it more precisely, one might ask, that makes people in Nelemnoye regard Sergei as a madman, in contrast to, for example, Irina, who also believes her grandson to be a returned deceased relative? I have already pointed out that although a child is considered the embodiment of a dead relative, he is also seen as a person in his own right. He carries within him an aspect of the dead, his soul, but is not the deceased person in any absolute sense. This is why one cannot simply interact with the child in the same way as one did with his namesake when that person was alive, but must adopt a double perspective. The problem with Sergei is that he relates to Vasili as if the boy was in fact his returned son. Thus, instead of moving between perspectives, Sergei has surrendered, so to speak, to the single

Figure 4.4 The burning of a deceased elderly man among the Chukchi

He apparently asked to be killed by his close relatives. The corpse is dressed in a special death garment made from reindeer skin and is burned together with a spear, tobacco purse and other things needed on the journey to the ancestor's realm.

perspective of "father." He thus conserves his past memory of his dead son in a manner that prevents him from recognizing that rebirth does not imply that he simply experiences the same person, but also—and most significantly—the return of someone other. Sergei, therefore, claims unity or identity between two persons, who are considered alike but not identical, and it is precisely this confusion of analogy with identity that makes him mad in the eyes of the community.

Among the Chukchi of Achayvayam, who still practice so-called "voluntary death" in which a person can request—even demand—to be killed at the hands of close relatives, it is quite common that this situation is brought about not simply as a result of old age and physical illness, but also by the trauma of the loss of a loved one (Willerslev 2009).

This was also noticed by the ethnologist Waldemar Bogoras (1904–9: 561) who, in his classic study of the Chukchi, wrote that "deep sorrow on account of the loss of some near friend must also be mentioned as a reason for voluntary death." Though I did not witness a case of voluntary death at first hand, I was told about an elderly man who, a few years before my arrival, had requested being killed, because he wanted to follow his dead wife to the afterworld. I was told that a young girl in the close family was considered to be the returned wife. However, the husband did not care about the girl and simply said "Oh! I miss my wife too much, how can I continue living? Take me to her." The others tried to persuade him to abandon the idea, but only for a while, since once a person has requested to die, there is no way back as the ancestors are expecting his arrival. Thus, the old man was soon after strangled.

The two stories reveal that although Siberian ideas about the return of dead relatives in newborn children can be seen as a particular cultural model for dealing with grief, which emphasizes the continuity rather than the finality of personal relationships, it does not always achieve the feelings of comfort intended. The relationship between the child and the bereaved is, in fact, a contingent process, which is inherently interactive and fundamentally risky. Everything, from the development of emotional bonds between the child and the bereaved to the latter's sense of moving from grief into contentment, depends on whether the parties involved can bring it off in their daily interaction with each other. There are strong emotions at stake, and many things can go wrong. The child may feel that his autonomy is being violated and therefore reject the identity ascribed to him, or the grief of the bereaved may be so overwhelming that the child is never actually allowed to fill the void left by the death of the beloved.

With these observations in mind, it seems that Freud was both right and wrong in his thesis of how the mourner facing loss may regress into a depressive state of melancholia. Freud was wrong with respect to his claim that a favorable progress of mourning must necessarily imply a detachment from the dead and an acceptance that acts of love can no longer be returned. The Siberian material points to the opposite, namely that by continuously investing in the lost love object, the bereaved may in fact fill the emotional void left by the deceased. Freud was right, however, in concluding that the memory of the lost person can become too dominant, which may result in the mourner's loss of the will to live. Why, we may ask, did Freud only develop a partial understanding of the nature of mourning and the process of grief-relief? I believe that it has to do with his inability to grasp the real nature of time, repetition and memory. This is what I will turn to discuss now.

Time, Repetition and Memory

The anthropological literature is crammed with typologies that crudely contrast the linear time of Western society with the repetitive time of indigenous ones. One example is Edmund Leach (2004 [1961]), who argues that primitive society is marked by two logically quite distinct "basic experiences" of time. One is founded upon the observation that certain natural phenomena repeat themselves, the other upon the observation that the individual person's life-course is irreversible and death inevitable. He goes on to suggest that the invariable strategy of primitive religious thought is to try to convince its practitioners that the kind of time we live in is along the lines suggested by the eternal return of life rather than on the irreversibility of death. In other words, ideas about repetitious time and immortality of the soul are, according to Leach, essentially religious fictions, whereas linear time is real or objective time.

But can we really trust this outline of so-called "primitive" thinking about time and can Leach's view on linear time as real or objective time really withstand serious philosophical scrutiny? I believe not, if we are to follow Deleuze, who is arguably considered the modern philosopher of time *par excellence*. I can only outline Deleuze's time theory here in bold strokes, but the point I hope to bring across is that although it flies right in the face of our commonsense understandings of time as a simple linear progression, it finds its creative equivalent among the Yukaghir and Chukchi.

Deleuze (1988) draws on his philosophical predecessor Henri Bergson's basic distinction between the "actual" and the "virtual." That something is *actual* principally means that it exists in the conventional sense of being empirically present—that it can be experienced, observed, and so on. The *virtual*, by contrast, is not physically given or presented, but can only be comprehended on a purely imaginary plane. This does not, however, make the virtual unreal, but quite the opposite. According to Deleuze, the force of the virtual is in an important sense more *real* than its actual manifestations. The virtual constitutes the primordial groundwork of the observable world's actual living manifestations—that is, their real starting-point or derivation.

This is also so with regard to time. We tend, for obvious reasons, to be focused on the present as the present is that dimension of time in which things literally fall into place—that is, time becomes actualized in the present as concrete and physical extensions in space. But it is this obsession with the present that mistakenly leads us to believe that past and present time is connected in a simple linear way. The true fabric of time is in fact virtual, which is not the present but the "non-present," that is, the whole continuum of time's past—what Deleuze (1994: 105) denotes the "pure past" and which Rune Nyord in this volume (Chapter 10) talks about as a "past that has never been present." The pure past, therefore, does not represent something that has been, but simply something that *is*, which links to and thus informs the present as its very precondition. To put it in the words of the Deleuzeian philosopher, Peter Hallward:

> The present is actual, the past is virtual. And for the same reason that the actual, despite its seemingly solidity, is in reality ephemeral and illusory, so too is the virtual, despite (or rather on account of) its immateriality and non-presence, the only true and lasting dimension of reality. (Hallward 2006: 33)

The pure or virtual past finds its matching counterpart in what Bergson (1988: 106) called "pure memory," which is the survival of memories. Time, in other words, is the absolute temporal movement informed by the past and retained in pure memory. Although the past is perfectly preserved in pure memory, we cannot access it in actuality. Pure memory only exists in the virtual dimension. Still, Deleuze (2000: 58) argues, when we truly remember something we do not reactivate a moment that was once present and now exists only as a mental trace of what it was once, but rather place ourselves directly in the past itself. In other words, we do not so much remember the past as we allow the past to think through us (Hallward 2006: 34).

If we turn our focus back to Siberia, there is an obvious overlapping between Deleuze's theory of time and the indigenous ideas about rebirth. As should be clear by now, to say that the present people are returned deceased ancestors is not to recall the tired cliché that Leach proposes for "primitive" peoples, namely that time repeats itself. If that were the lesson of repetitious time, it would have nothing to teach us. Everything returns, yes; everything recurs. But what recurs does not do so in the form of actualized identities but in the form of the virtual that constitutes those identities. Indeed, this is why the Siberian peoples insist that it is the invisible, ethereal and ghostlike aspect of the person—that is, the soul—that recurs, not his body or the person *in toto*. It is the soul that belongs to virtuality. The same goes for the person's "name" which, according to native sayings, is equated with the soul; what allows us to consider two things as identical is exactly the signifier or name. As philosopher Bruce Fink (1995: 224) writes: "You can only step in the same river twice because you have a word or name for it – the Swanee River, for example." At the level of the signifier, then, repetition implies the return of something that in actuality would be different the second time, but for the name.

Accordingly, it is only on the plane of virtuality that we can talk about repetition of the same. On the plane of actuality, repetition consists of the return, not of the same, but of the different—the return of something other. As Deleuze puts it: "It is not being that returns but rather the returning itself that constitutes being … In other words, identity in the return does not describe the nature of that which returns but, on the contrary, the fact of returning for that which differs" (Deleuze

1983: 48). What returns, in other words, are not the identities of people and things that are actualized in the present; what returns is the virtuality—the soul—that lies behind and within those identities. It is exactly for this reason that the bereaved in Siberia cannot simply interact with the incarnation of a dead person as if he or she was in actuality the dead person, but must adopt what I have denoted a double perspective: one perspective is placed in the past and links to the memory of the dead, to virtuality; the other is placed in present and links to the living, to actuality. The two perspectives are mutually exclusive, yet jointly they form parts of the same reality. Thus, the reality of rebirth is comprised of both the virtual and the actual; just as the living person is a creature of both the pure past and the actual present. Indeed, this is also the critical idea of Deleuze (1994), when he writes: "What is produced, the absolute new itself, is in turn nothing but repetition" (Deleuze 1994: 90).

In one sense this is a far cry from Freud's theory. Freud's understanding of time as a linear progression in which the past goes before the present prevents him from acknowledging the virtual nature of time. As a result, in Freud's model the barrier between the past and the present, between the living and the dead, is essentially impermeable. For Freud, "reality-testing" implies that the past lives of the dead do not really exist, except in the unstable mind of the melancholic patient, and for this reason, "all libido must be withdrawn from its attachments to that object" (Freud 2009: 255). However, as we have just seen, the virtual past exerts its influence directly through its capacity to link to and thereby inform the present. So, although the past lives a kind of shadowy and fleeting existence, it still *is*—it is *real*.

Freud's narrow understanding of reality as simply including the actuality of the present also leads him to develop an inadequate theory of repetition. Freud conceives of repetition only as repetition of the same, one that does not give way to difference but denies the very possibility of difference. Indeed, this is what makes him conclude that the "repulsion to repeat" is equivalent to the death drive as an inert desire to "restore an earlier state of things" (Freud 1961: 32). In other words, for Freud the point of all repetition is to reach the same, previously existing, state of affairs. As a result each situation forgoes its novelty and is merely repeated on the model of the previous. Freud thus fails to see that repetition harbors the virtual power to transform what it repeats. Indeed, this is the very principle on which Siberian rebirth as grief-relief is based. Rather than stopping the repetition, the mourner continues to repeat, constantly investing in the lost love object, exactly because repetition always leads to the return of something else, something new.

Still—and this is a key point—implicit in Freud's somewhat crude and narrow understanding of time and repetition lies a revolutionary insight, namely that melancholic people are those for whom the past carries more weight than the present can bear; those who cannot move beyond memory. For them the present is simply a continued confrontation with the past. This is a lived time that does not stop orienting itself toward the past, thus deforming the present in order to make it a copy of the past. In other words, what Freud discovered was that the virtuality of time along with the memory it entails has the capacity to make people ill, even killing them. This brings me to the final task of this chapter, to revise Freud's meaning of the death drive. The gist of my interpretation is to understand the death drive not as a departure from Freud's view, but as a kind of radicalization of it.

The Meaning of the Death Drive

I have already described how Freud clung fast to the concept of repetition as a kind of indispensable key to unlocking the mystery of the death drive, yet without ever fully unlocking it. In his last major article, "Analysis Terminable and Interminable" (1955 [1937]), he insisted more strenuously than ever on the necessity of the dual drive hypothesis, claiming that only by "the two primal instincts – *Eros* and the death instinct – never by one or the other alone, can we explain the rich multiplicity of the phenomena of life" (1955: 243). Freud thus asserted a paradoxical connection between the death drive and the realization of life's full vitality, but he was incapable of fully developing the link. The reason, I believe, and which I have already touched upon, lies with his inadequate understanding of the nature of time and repetition. For Freud, as we have seen, repetition simply meant the repetition of the same on the level of actuality. However, with Deleuze, we have seen that repetition belongs to virtuality. The virtual past is free to circulate in whatever presents it may press itself upon, creating endless diversity in and through every repetition. Indeed, for Deleuze, this is the eternal nature of *creation*, which is the very source of life itself. However, Freud discovered another side to this perpetual renewal of life, namely its destructive side, what he called *Thanatos*, meaning "death" in Greek. While Deleuze finds that the pure past is the very condition of life's creativity, Freud made the important discovery that people who get trapped in the past—that is, in memory—transform into the very negation of life, taking on such hideous symptoms as melancholic depression and suicidal desire. In short, Freud saw that life as creation also works in the service of death.

I have now reached the crux of my argument. Perhaps we are in a position to explain the meaning of the death drive? For we may ask: what is it that returns eternally, which has the power to dance in and out of potency and despair, creation and destruction? It can paradoxically only be the very opposite of death, namely *life*. Not simply life in its actual present form of *Eros*, but in an uncanny *excess* of life, an indestructible insistence of the libido that persists beyond the biological cycle of life and death. So, the Freudian death drive is actually the name for its very opposite, namely life as a virtuality of infinite becoming. I largely derive this point from the cultural critic Slavoj Zizek (2009), who draws a revealing analogy between the death drive and the monstrous creature in Ridley Scott's film *Alien*. The alien's form of life is, as Zizek writes, not merely life, it is life *as such*: "It is … a nightmare embodiment of the natural realm understood as utterly subordinate to, utterly exhausted by, the twinned Darwinian drives to survive and reproduce" (Zizek 2009: 118). In other words, Scott's alien is in essence the libido, indestructible and immortal. And so, the death drive is the name for eternal life itself, the compulsion to engage in the endless circular movement of expanded self-reproduction and survival. This is why Zizek is quite right in arguing that the movement of the death drive must be the "weird logic of the curved space in which the shortest distance between two points is not a straight line, but a curve: The [death] drive 'knows' that the shortest way to attain its aim is to circulate around" (Zizek 2006: 61).

Although the Yukaghir and Chukchi do not say it in so many words, this must be their message too, when they talk about the eternal recurrence of the soul, ignoring all hindrances and repeating the same circle again and again. Is must also be what they have in mind when talking about the soul as indivisible, indestructible and immortal and when they describe how it can make its home in a multiplicity of bodies. Indeed, I venture to suggest that this blind, indestructible insistence of pure life must be what Freud called the death drive.

References

Aslan, Carlos M. 2009. "Mourning and Melancholia: A Freudian Metapsychological Updating," in Leticia Glocer Fiorini, Thierry Bokanowski and Sergio Lewkowicz (eds) *On Freud's Mourning and Melancholia*, pp. 162–78. London: Karnac Books.

Bergson, Henri. 1988. *Matter and Memory*, trans. Nancy Margaret Paul and Scott Palmer. New York: Zone Books.

Bodenhorn, B. 1989. "I'm Not the Great Hunter; My Wife Is: Inupiat and Anthropological Models of Gender," *Etudes/Inuit/Studies* 14(1–2): 55–74.

Bogoras, Waldemar. 1904–9. *The Chukchee*, edited by Franz Boas. Memoir of the American Museum of Natural History. New York: American Museum of Natural History.

Born, Georgina. 1998. "Anthropology, Kleinian Psychoanalysis, and the Subject in Culture," *American Ethnologist* 100(2): 373–86.

Boothby, Richard. 1991. *Death and Desire: Psychoanalytic Theory in Lacan's Return to Freud*. New York and London: Routledge.

Cohen, Josh. 2005. *How to Read Freud*. London: Granta Books.

Deleuze, Gilles. 1983. *Nietzsche and Philosophy*, trans. Hugh Tomlinson. Minneapolis: University of Minnesota Press.

Deleuze, Gilles. 1988. *Bergsonism*, trans. H. Tomlinson and B. Habberjam. New York: Zone.

Deleuze, Gilles. 1994. *Difference and Repetition*, trans. P. Patton. New York: Columbia University Press.

Deleuze, Gilles. 2000. *Proust and Signs*, trans. Richard Howard. London: Continuum.

Devereux, George. 1978a. "Two Types of Model Personality Models," in *Ethnopsychoanalysis: Psychoanalysis and Anthropology as Contemporary Frames of Reference*. Berkeley and London: University of California Press.

Devereux, George. 1978b. "Ethnopsychoanalytic Reflections on the Notion of Kinship," in *Ethnopsychoanalysis: Psychoanalysis and Anthropology as Contemporary Frames of Reference*. Berkeley and London: University of California Press.

Fink, Bruce 1995. "The Real Cause of Repetition," in Richard Feldstein, Bruce Fink and Marie Jaanus (eds.) *Reading Seminar XI: Lacan's Four Fundamental Concepts of Psychoanalysis*, pp. 223-229. New York: State University of New York Press.

Freud, Sigmund. 1955 [1937]. *The Standard Edition of the Complete Psychological Works of Sigmund Freud*, 24 vols. Edited and translated by J. Strachey and A. Freud et al. London: Hogarth Press & The Institute of Psychoanalysis.

Freud, Sigmund. 1961. *Beyond the Pleasure Principles*, trans. J. Strachey. New York: W.W. Norton.

Freud, Sigmund. 2009 [1917]. "Mourning and Melancholia," in Leticia Glocer Fiorini, Thierry Bokanowski and Sergio Lewkowicz (eds) *On Freud's Mourning and Melancholia*, pp. 19–32. London: Karnac Books.

Gellner, Ernest. 1985. *The Psychoanalytic Movement*. London: Paladin.

Guemple, Lee. 1994. "The Inuit Cycle of Spirits," in Antonia Mills and Richard Slobodin (eds) *Amerindian Rebirth: Reincarnation Belief among North American Indians and Inuit*, pp. 107–22. Toronto: University of Toronto Press.

Hallward, Peter. 2006. *Out of This World: Deleuze and the Philosophy of Creation*. London: Verso.

Heald, Suzette, Deluz, Ariane and Jacopin, Pierre-Yves. 1994. "Introduction," in Suzette Heald and Ariane Deluz (eds) *Anthropology and Psychoanalysis*, pp. 1–26. London and New York: Routledge.

Hertz, Robert. 1960 [1907]. "A Contribution to the Study of the Collective Representation of Death," in *Death and the Right Hand*, trans. R. Needham and C. Needham, pp. 27–86. Glencoe, IL: The Free Press.

Holbraad, Martin. 2007. "The Power of Powder: Multiplicity and Motion in the Divinatory Cosmology of Cuban Ifá (or *Mana*, Again)," in Amiria Henare, Martin Holbraad and Sari Wastell (eds) *Thinking through Things: Theorising Artefacts Ethnographically*, pp. 189–225. London: Routledge.

Jochelson, Waldemar. 1926. *The Yukaghir and the Yukaghized Tungus*, edited by Franz Boas New York: American Museum of Natural History.

Leach, Edmund. 2004 [1961]. "Two Essays Concerning the Symbolic Representation of Time," in *Rethinking Anthropology*, pp. 124–36. Oxford and New York: Berg.

LeVine, Robert A. 2001. "Culture and Personality Studies, 1918–1960: Myth and History," *Journal of Personality* 69(6): 803–18.

Malinowski, Bronislaw. 1929. *The Sexual Life of Savages in North-Western Melanesia*. London: Routledge.

Mills, Antonia M. and Slobodin, Richard (eds) 1994. *Amerindian Rebirth: Reincarnation Belief Among North American Indians and Inuit*. Toronto and London: University of Toronto Press.

Nuttall, Mark. 1994. "The Name Never Dies: Greenland Inuit Ideas of the Person," in Antonia Mills and Richard Slobodin (eds) *Amerindian Rebirth: Reincarnation Belief among North American Indians and Inuit*, pp. 123–36. Toronto: University of Toronto Press.

Obeyesekere, Gananath. 2002. *Imagining Karma: Ethical Transformation in Amerindian, Buddhist, and Greek Rebirth*. Berkeley: University of California Press.

Paul, Robert A. 1980. "Symbol Interpretation in Psychoanalysis and Anthropology," *Ethos* 8(4): 286–94.

Rivers, W.H.R. 2001 [1932]. *Conflict and Dream*. Kegan Paul, Trench, Trubner & Co., Ltd.

Schrempp, Gregory. 1992. *Magical Arrows: the Maori, the Greeks and the Folklore of the Universe*. Madison: University of Wisconsin Press.

Seligman, C.G. 1924. "Anthropology and Psychology: A Study of some Points of Contact," *Journal of the Royal Anthropological Institute of Great Britain and Ireland* 54: 13–46.

Spiro, Melford E. 1982. *Oedipus in the Trobriands*. London and Chicago: University of Cihicago Press.

Strathern, Marilyn. 1988. *The Gender of the Gift: Problems with Women and Problems with Society in Melanesia*. Berkeley: University of California Press.

Tähkä, Veikko. 1984. "Dealing with Object Loss," *Scandinavian Psychoanalytic Review* 7: 13–33.

Vitebsky, Piers. 1993. *Dialogues with the Dead: The Discussion of Mortality among the Sora of Eastern India*. Cambridge: Cambridge University Press.

Vitebsky, Piers. 2005. *Reindeer People: Living with Animals and Spirits in Siberia*. Great Britain: Houghton Mifflin Harcourt.

Whiting, John W.M. 1953. *Child Training and Personality: A Cross-cultural Study*. New Haven: Yale University Press.

Willerslev, Rane. 2007. *Soul Hunters: Hunting, Animism, and Personhood among the Siberian Yukaghirs*. Berkeley: University of California Press.

Willerslev, Rane. 2009. "The Optimal Sacrifice: A Study of Voluntary Death among the Siberian Chukchi," *American Ethnologist* 36: 693–704.

Willerslev, Rane. 2012. *On the Run in Siberia*. Minneapolis: University of Minnesota Press.

Zizek, Slavoj. 2006. *Lacan*. London: Granta Books.

Zizek, Slavoj. 2009. *The Parallax View*. Cambridge and London: MIT Press.

Chapter 5

Sharing Death: Conceptions of Time at a Danish Online Memorial Site

Dorthe Refslund Christensen and Kjetil Sandvik

Introduction: On Temporal and Cosmological Collapses

In his book, *Event and World* (2009), the French philosopher Claude Romano establishes a phenomenology of events and theorizes on the human condition within a particular framework:

> [A]n event … illuminates its own context, rather than in any way receiving its meaning from it. It is not a consequence of this context that could be explained in light of preexisting possibilities, but it reconfigures the possibilities that precede it and signals the advent of a new world for an advenant.[1] This is not to say that the former world entirely disappears as such, but that its *meaning* is so radically modified, the totality of projections and finalities that dwelled in it and conferred a signifying structure on it are so altered, that strictly speaking it is no longer the *same* world: the occurring of such an event renders the former world insignificant, since this event can no longer be understood in light of that world's context. (Romano 2009: 38f.)

The death of an infant is such an event. When parents are expecting, the child is usually awaited with hope and plans and reflections on how to share the life and future with this new child. As such the child opens the parents' perspective to a new future, a span of time or potentiality, a certain narrative—that is, being parents, raising a child. When a child is stillborn or dies at a very young age, this event not only turns off all hopes for this particular child and all good things envisioned for the child and his or her family, the world itself changes radically.

At the Danish website Mindet.dk parents[2] engage in ritualizations through which they re-situate themselves in the world through performing their grief

1 An "advenant" is Romano's term "for the human being insofar as something happens to him and … he is open to events" (Romano 2009: 21). As such this is the core concept of Romano's "evential hermeneutics" (2009).

2 Mindet.dk is an online forum for all those who experience loss and grief in relation to the death of spouses, parents, grandparents and friends. However, the work we are presenting focuses solely on the ritualizations performed by parents of stillborns and infants

and loss, re-relating themselves to others and renewing their acquaintance with themselves and the world. These ritualizations are carried out through narratives and performances dealing with different aspects of the loss and are not about "translating" the everyday experience into, for instance, a religious realm. Rather "the work of ritual … involves developing repertoires that operate in complex interplay with the world of everyday experience" (Seligman et al. 2008: 29) and, eventually, through time and repetition, lead to re-experiencing life.

The ritual work performed at Mindet.dk is complex and diverse. However, in this chapter we apply a "time" perspective as our analytical take on these ritual practices and we suggest, as our point of departure, that with the death of very young children time itself is utterly distorted and the future of the parents[3] disappears while the presence/present is folded into the past, leaving the parents in some kind of *non-time*. We analyze how parents on Mindet.dk engage in ritualizations in which "timework" plays a significant role and how these practices aim at re-enrolling the parents in a temporal reality where past, present and future exist and can be appropriated, negotiated and, eventually, re-inhabited.

Mindet.dk: A Brief Presentation

Over the past 10–15 years a variety of websites have sprung up on the Internet that facilitate the process of mourning and commemoration for people who have lost loved ones (children, lovers, siblings, parents, friends), from general social network sites like Facebook (which—among a lot of other things—harbors both groups and profiles for mourning and grieving) to designated websites like, for example, To Heaven With Love. Mindet.dk belongs to this last category of websites (*mindet* means memory). On this website participants perform their grief by designing online memory spaces for their loved one(s), displaying photographs, poetry, stories and expressions of grief and longing. They mark red letter days and difficult times by lighting online candles for their child and they express empathy for others by lighting the same kind of candles for other people's loved ones. They share their personal experiences in different chat rooms and the website offers services such as a calendar displaying anniversaries and different guestbook facilities attached to each memory profile.

---

since these practices are special from a time perspective because of the very short lives of the deceased.

3 At Mindet.dk it might seem as if it is the mothers who are the active parts, whether writing "we" or "I." In fact the mothers are the visible mourners and we have failed to find a father being responsible for a memory profile. However we do not want, at this stage of our research, to make a point about the mother suffering the most. We observe that the mother seems to be the one taking on the emotional labor in the loss but we do not have material that can throw light on the absence of fathers.

Mindet.dk was established in 2000 by a private person (Christian, no last name) as a "place for those we got, loved, and lost." In 2008 the website and its around 230 memory profiles were taken over by the Danish newspaper *Kristeligt Dagblad* (*The Christian Daily*), which is still running the website with 920 memory profiles (April 2012). According to Wikipedia, about 2,000 people visit the website every day to commemorate their lost ones and between 100 and 200 virtual candles are lit during a regular day.[4]

Mindet.dk enables the users to create memory profiles for their lost loved ones. These consist of a front page with the possibility to display pictures, texts and animations and the profile can be customized by choosing among predefined background pictures or by uploading the user's own picture/graphic material. Each profile may have various archives containing, for example, diaries or pictures and stories from the funeral, the burial site and so on and each profile comes with a calendar for displaying red letter days (typically, birth and death dates) and a guestbook for others to post greetings and comments. Beside these functionalities, the website includes a "Light-a-Candle" function which may be used for short messages, such as the mother lighting a candle at bedtime for her dead child or a relative or another member of the Mindet.dk community sending a greeting on a red letter day. Mindet.dk also includes a "Forum" for debate on a variety of topics dealing with loss and grief, and this forum plays an important role in constituting Mindet.dk not just as a place for individuals' processes of grieving, but also a place for peer-to-peer compassion and caring between people sharing similar experiences and going through the same processes of getting life back together. In all of these practices "doing time" is a central way of "doing grief."

Mindet.dk seems to fill a void for having marked or designated places for mourning and bereavement (after the funeral and in addition to the graveyard), not the least in the case with the loss of infants or with stillborns: deaths which more than others leave the bereaved asking "why" and a lengthy process of mourning and reworking the meaning of life. These types of deaths, which are in focus of this chapter, are overrepresented at Mindet.dk compared to annual death statistics: out of the approximately 55,000 people who die every year in Denmark, less than 1 percent are under 10 years of age, while at Mindet.dk this same age group is the subject of around 40 percent of the individual memory profiles (and at the forum the largest number of discussion threads are on dead children and stillborns). The quality of an online community shows in this respect; this is where users with the same experiences can meet. As stated by a mother of a stillborn daughter: "Mindet.dk has helped me finding someone to talk to – someone in the same situation as me. It is good to know that I am not the only

4 Obviously, this Wikipedia entry may have been written by the Mindet.dk owners themselves. Still, judging by the number of memory profiles, entries in guestbooks and candles being lit every day, this seems like a fair estimate.

one to feel loss and sorrow, and it is good to hear how others feel and how far they have come in their process of grief and be able to help them."[5]

Mindet.dk forms a genuine community where individuals are offered a ritualized (marked) space for their individual performances of grief *and* at the same time are given the possibility of performing empathy toward others mourning their own loved ones. It becomes evident that media, in this part of the world, have a great influence on life practices such as grieving and mourning: the bereaved are offered a medium and a technology that are familiar (computer, the Internet and chat rooms) and which provide the opportunity to express grief within one's own cultural and social spaces as opposed to offline mourning spaces where the language is set aside from ordinary life (for example, in the church). By enrolling themselves in this "editable" community the bereaved commit themselves to an ongoing communication with the person they lost and with other people in the same situation as themselves. And this continued dialogic practice accentuates the feeling of still being in close contact with the deceased and being allowed to perform grief without any time limits: "In short, Mindet.dk is a place where you get to feel sorrow indefinitely; everyone here has experienced great losses, you are not alone … and most important: [with] the great compassion expressed by everyone, you feel that you are really being taken care of."[6] So, Mindet.dk is a spatial, a temporal and a ritual construction and these aspects are, of course, interrelated in the practices found at the site.

Notably, Mindet.dk is perceived as a place, and as we will show, the materiality of the system which connects to the website as a signifying structure relates both to the temporal and spatial dimension. The materiality of Mindet.dk is characterized by having marked entrances or individual links leading to segregated "locations" (the individual memory profiles) separated from the rest of the site (and the web). These designated places include a marked difference between an *inside* and an *outside* defined by personalized design such as background texture, imagery and colors, use of text, pictures, drawings and so on. Furthermore, the signifying *inside* is also characterized by a specific set of narratives and performances,[7] which we will be analyzing throughout this chapter. The performative aspect is important in that Mindet.dk is not just a place for performance—of grief and mourning and (re-)production of memories—it is also a place which comes into being through performative acts. Thus the media-created space is also functioning as a creative tool "writing" ritual practices and relationships; this makes Mindet.dk both a place for individual mourning and a community for mourners.

---

5 http://mindet.dk/node/65902, accessed April 16, 2012.

6 http://mindet.dk/node/65902, accessed April 16, 2012.

7 The connection between the narrative and the performative may be described this way: while a narrative is the telling of (past, present, future) actions and events taking place, a performative is the defining of the actions and events that are making them come into being.

Time and Narrative

In the following we analyze the complex strategies implied in these practices on Mindet.dk through three different sets of theoretical thinking about time that provide a framework for our analysis: (1) the reflections on narrative and time put forward by philosopher Paul Ricoeur (1980) in his extensive writings in the 1980s;[8] (2) the concept of *timework* suggested by sociologist Michael G. Flaherty in his book *The Textures of Time: Agency and Temporal Experiences* (2011); and (3) the concept of *chronotope* introduced by literary theorist Mikhail M. Bakhtin in his essay on "Form of Time and Chronotope in the Novel" (Bakhtin 1981). Basically, we suggest considering the mourning practices at Mindet.dk as special ways of doing time (Flaherty 2011). We are interested in how mourners engage in different kinds of timework and how this structures their ritual performances of grief. Ricoeur seems to implicate both the performative and the ritual; however he does not describe either very extensively. By stressing the performative character of timework we add a new dimension to Ricoeur's work and his concept of *narrative time* by pointing to the performative and ritual character of time.

Paul Ricoeur understands narratives as major life metaphors employed by people: narratives are devices for understanding both our collective history (who we are, what we stand for) and our individual, autobiographical stories. Narratives enable us to make sense of the various occurrences we encounter in our lives, to unfold our very existence and the conditions and possibilities we live under. As such, we understand our lives by transforming them into narrative entities structured by well-known elements, such as beginnings and ends, organizing whatever we encounter into chains of events, into "plots": "By plot I mean the intelligible whole that governs a succession of events in any story … A story is *made out of* events to the extent that plot *makes* events *into* a story. The plot, therefore, places us at the crossing point of temporality and narrativity" (Ricoeur 1980: 171). The very logic of the plot is to produce meaning: "Every narrative combines two dimensions in various proportions, one 'chronological' and the other 'nonchronological'. The first may be called the 'episodic dimension', which characterizes the story as made out of events. The second is the 'configurational dimension', according to which the plot construes significant wholes out of scattered events" (Ricoeur 1980: 178).

The narrative timework we use in order to structure and make sense of our lives is based on the same principles. But the configurational mode is not just for organizing events of the past but also, in a subjunctive way, for possible future events, which is summed up in our "sense of an ending" (Ricoeur 1980: 179): "By reading the end in the beginning and the beginning in the end, we learn also to read time itself backward, as the recapitulating of the initial conditions of a course of action in its terminal consequences. Memory, accordingly, *repeats* the course of

8 Ricoeur 1984 [1983], 1985 [1984], 1988 [1985].

events according to an order that is the counterpart of time as 'stretching-along' between a beginning and an end" (Ricoeur 1980: 180).

As we will demonstrate, the logics of recollecting and repeating events are at play in the performative ritual practices at Mindet.dk to an extensive degree. The cyclic actions performed when it comes to marking red letter days such as birthdays and death days or just the almost automated lighting of candles at bedtime, demonstrates the need for creating narrative structures which ensure that the mourners may develop and keep alive memories and relations to the dead child. Therefore the "stretching of time" is not something that takes place between a beginning and an end, it is a stretching of time to ensure duration (Flaherty 2011), to make the moment last, as becomes evident in this inscription from a children's grave: "You gave us a day which lasts forever."[9]

We understand our lives through the use of narratives and this operation also connects to our temporality, our memories, presence and expectations and thus to the basic human condition of being tied to/part of a plural unity of past, present and future. In other words, time is life being narrated, and the narrative *is* time, its core components of beginning, middle and end are temporal concepts.

According to Ricoeur, time and spatiality are interlinked in ways similar to Bakhtin's chronotope: we are situated "in time" and this "within-time-ness," Ricoeur claims, using a Heideggerian concept, "possesses its own specific features which are not reducible to be representation of linear time, a neutral series of abstract instants" (Ricoeur 1980: 173). So what Ricoeur is pointing at is, in fact, that our lives consist of a number of diverse, fragmented and heterogenous experiences but when this collection is recollected and turned into a narrative that also gives it a specific direction then time itself becomes narrative. Time becomes a path or route organizing the abovementioned occurrences we may encounter in our lives and thus metaphorizing life as a journey through a course of meaning-*given* events.[10]

Our point in this chapter, adding to Ricoeur, is that when this narrative is rooted in or expressed through performative practices we can talk about *performative time*. As we see it, performative space-time is established through action rather than narration. It addresses our effort to create and make time thus dealing with time in proactive ways, which is an important part of the ritual practices at Mindet.dk such as the lighting of candles, celebrating birthdays and so on. Through such

---

9 Albert Arve Versterbæk's grave, Nordre Kirkegård, Aarhus, Denmark.

10 Cultural studies analyst John Storey (2003) suggests that humans engage in their identity work by organizing the narrative processes in such a way that the past ("roots") is giving sense to the present in which we stand and, not the least, is guiding us for the "routes" we plan to take in the future. His basic point is that roots do not refer to an objective past that we can recall as memory in a banal sense but rather, that memories are something we produce in order to make sense of what is going on and of the plans and strategies we engage in. So roots are not solely recalled, they are "produced." The future is the time onto which we project our hopes and efforts.

practices time is negotiated, re-created and re-appropriated. The death of a child represents the collapse of time and the process of mourning is about taking time back or re-installing differentiated time and thus re-installing or re-creating meaningful time.

According to Ricoeur, time is not just passing by, it is something that we make, explicitly evident in expressions such as "having time to," "taking time to," "wasting time" and so on (1980: 173), and the concept "now," for instance, does not necessarily imply a chronological or linear thinking stating present to be different from past per se but might instead be understood with Heidegger and his idea that "Saying now … is the discursive articulation of a *making-present*" (from Heidegger's *Being and Time*, part 2 1962: 469, cited in Riceour 1980: 173). This performative time-making may be used for describing a key component in the timework at play at Mindet.dk, namely that life, or the present, is re-established as a meaning-making entity by ways of re-working and re-telling the past. This is done in order to re-invent the future, thus transforming it into something other than a paralyzing factor, and by applying the subjunctive aspects of ritualization, inscribing the loss and memories of the child and its potential future as a bearable part of the life to be.

Performing Time: On Timework and Ritualizations

According to sociologist Michael G. Flaherty, timework can be referred to as an "intrapersonal and interpersonal effort directed toward provoking or preventing various temporal experiences" implicating "the agentic micromanagement of one's own involvement with self and situation" (Flaherty 2011: 2).[11] Flaherty analyses "to what extent and in which ways … individuals purposefully construct lines of activity or social situations in order to create or inhibit diverse forms of temporal experience" (2011: 11). Basically we claim that that mourning ritualizations at Mindet.dk can be productively analyzed as timework and that by engaging in different kinds of timework at Mindet.dk, the mourners are eventually transforming past into a time they live *with* as opposed to the past being a time they live *in*. The dead child is transformed from "the child that did not live," which caused a paralyzing time-space of shattered hopes and broken dreams—the family that never came to be, the parenthood that never happened—to a character or agent living its own life in parallel (and with connections to) the new life, the new narrative, of the parents. The parents do this, as we will demonstrate, by narrating the life and death of the lost child and by performing parenthood; both of which are organized in close relation to time; thus narrating life is the process of doing time in context.

11 According to Flaherty, a distinction can be made between six different analytical categories: duration, frequency, succession/sequencing, (optimal) timing, allocation (distribution) and taking time (Flaherty 2011: 12). We do not use these categories systematically in our analysis, although we are inspired by them.

Flaherty points out that rather than looking at time as "situationally given … we must see that the individual strives to control or customize how time is experienced" (2011: 132). The parents' timework leads to a re-ordering of their world that has been so deeply distorted by the death of their child. Through the ritualized performances the parents re-organize their own signifying structures and the re-construction of the parents' cosmological and temporal order works to adjust the parents to the narrative time of the social world and enable the child to persist in existing as a socio-cultural being. One of the reasons why the memories of stillborns are so interesting in relation to a time perspective is that these children have almost had no time among the living and a crucial part of the narratives and performances of the parents seems to be to compensate for this in order to ensure that the child is remembered and treasured. In this sense, timeworking at Mindet.dk resembles ritual as described by Seligman et al. (2008) and it seems inevitable that the timework at Mindet.dk be dealt with as a ritualization that has a sense of play to it, not in terms of having fun, obviously, but rather as a kind of ritual performance that has a strong sense of experimentation and negotiation to it.[12]

Adam B. Seligman, Robert P. Weller, Michael J. Puett and Bennet Simon in their book *Ritual and its Consequences: An Essay on the Limits of Sincerity* (2008) argue that what ritualizations basically do is to create a subjunctive, an "as if'" or "could-be" universe (Seligman et al. 2008: 7ff.) and that this shared subjunctiveness is not restricted to the religious realm but is rather a fundamental way of negotiating our very existence in the world. In continuation of these ideas we point to ritualization as a performative modus that can both point to certain aspects of life and create modifications, differentiations and demarcations in the socio-cultural world and, at the same time, negotiate the very premises of the social while performing itself within an experimenting and meta-reflexive framework. So even though ritualization might be about meaning it is not about expressing or acting out an already fixated meaning—rather it depends on the experimentation with meaning itself and the idea of the possibility of there being any meaning after the event of the child's death. Just as Romano points to the total transformational character of the event (as quoted above), one might say that one of the crucial ritual negotiations we find at Mindet.dk is whether or not new signifying structures can at all be found in the world after this event.

12 Even though, as Flaherty points out, timework is object to the creativity, routine and choices of the individual performer and thus a result of individual effort, time is performed along the norms of society and in order to make, in our case, death manageable along such norms: "human beings make their own temporal experience" but they do not make it just as they please; "they do not make it under circumstances [utterly] chosen by themselves, but under circumstances directly found, given, and transmitted from the past" (Flaherty 2011: 134, with a reference to Marx). In other words, the timeworker quotes socio-cultural matrices that are present in his or her world not by simply repeating them but by creatively performing customized versions of it in performative improvisations (cf. Ingold and Hallam 2007).

This points further to ritual as a social or relational practice. Ritualizations might be performed individually but they deal with the world and with being in the world as one human being among others. In ritual, humans reach out in both abstract and concrete form. This is crucial at Mindet.dk where the experiments with signifying structures and the social roles of the parents (how to move on, how to face friends and family, how to have fun with colleagues at work, how to be a parent with a dead child) are constantly mirroring other parents.

Anthropologist Catherine Bell (1997) has defined ritualization as "the simple imperative to *do* something in such a way that the doing itself gives the acts a special or privileged status" (Bell 1997: 168). Ritual points to itself in a way that lifts it above other socio-cultural practices. Ritual is not only doing something at random but doing something believed to be important or crucial in a staged way. This is in line with the work of anthropologists Caroline Humphrey and James Laidlaw (1994) when they point to ritualizations as acts that are stipulated and carried out within a certain attitude or modus of the ritual performer that makes the performances within this framework more important than daily life activities. This approach stresses two characteristics about ritual: that a ritual can be any act ascribed with a certain aura of importance and, secondly, that a ritualization is performed within a certain framing (Sjørslev 2007; Bateson 2000) that sets the practice aside from other practices. This might imply certain attitudes and gestures used as demarcations that separate "what is going on" from mundane activities in ordinary life and/or the ritual space might be a certain physical and/or virtual location. In fact, we suggest that Mindet.dk consists of a designated time-space for performing ritualized practices of mourning and commemoration which may be understood if we apply the concept of the chronotope as developed by Bakhtin.

Time and Space

Mikhail Bakhtin conceives the concept of *chronotope* as the intrinsic connectedness of temporal and spatial relationships that are artistically expressed in literature (Bakhtin 1981). Although we are clearly not preoccupied with an artistic expression in literature, we find the chronotope useful to denote the infrangible interrelations of time and space that we consider to be the core characteristic of all the ritualizations performed at Mindet.dk. We see the chronotope as a basic quality of ritual itself occupying some qualitative differences of time-space relations in the ritualized space-time at Mindet.dk, compared to the mourners' lives outside of the Mindet.dk practices.

The interconnectedness of time and space at Mindet.dk is our way of *framing* Mindet.dk's ritual practices and the social community-building accompanying some of these ritualizations (Sjørslev 2007). In this perspective Mindet.dk is a chronotopian system consisting of certain ritual practices *doing time* differently than in the sociality outside of Mindet.dk. The performative ritualization strategies at work at Mindet.dk are dependent on demarcations of space and time, that is, the making of a designated space and time for performing grief which thereby also segregates it from other aspects of social life ("in here I can spend time with my 'angel', outside I have to play the role as e.g. an employee").[13] According to Seligman et al. (2008), it is exactly the difference between life and ritual that makes ritual strong and desirable as a performative modus (Seligman et al. 2008: 29f.). By logging in, individuals/people step into the designated space-time in which they can "legitimately" spend time being preoccupied with their loss and their reflective process of mourning. In this sense, Mindet.dk, because of its chronotopical qualities, serves as an intensification device, a time-place for becoming immersed and focused.

We suggest that the chronotope serves as a "living metaphor" in a Ricoeurian sense, being a type of metaphor that enables us to navigate and make sense of our lives.[14] Thus the chronotope also possesses a certain kind of materiality: it informs and devises concrete life strategies. The chronotopical system at play at Mindet.dk has bearings on the (everyday) life of the mourners. The chronotope enables a structuring of time in ways which may produce meaning, focusing on significance rather than chronology. When we study the timework practices at Mindet.dk it becomes evident how these ways of creating time, that is, organizing and manipulating events into meaning-producing structures (plots, according to Ricoeur), are put to work; particularly when analyzing the subjunctive aspects of performative and ritual practice at work, we come face to face with the extra-ordinary logics of fantasies and dreams.

Performing Signifying Structures: An Analysis of Narrativity and Time at Mindet.dk

How does Mindet.dk function as an organizing device for the bereaved? How can time be reclaimed as meaningful? How can the tragic events of losing a child be plotted (cf. Brooks 1984) in new ways in order to produce narratives (cf. Ricoeur 1980) which may function as signifying structures (cf. Romano 2009). Due to the death of the child the parents at Mindet.dk have changed from being *parents to be* to *parents not to be*. Their main narrative of preparing for *the parenthood to be* has been broken into bits and pieces and this causes a narrative collapse which may best be characterized, we suggest, as a situation where the direction of time is reversed. The death of the child becomes an event of the past and of such strong significance, that the future dissolves as a meaningful entity and the present may be described as folded into the past, because of the all-absorbing emotions of loss and grief connected to the bereavement.

---

13 In fact the demarcation seems to be crucial in order to segregate the practices at Mindet.dk from everyday life. The demarcation is an important part of a strategy for moving on.

14 Ricoeur (1975).

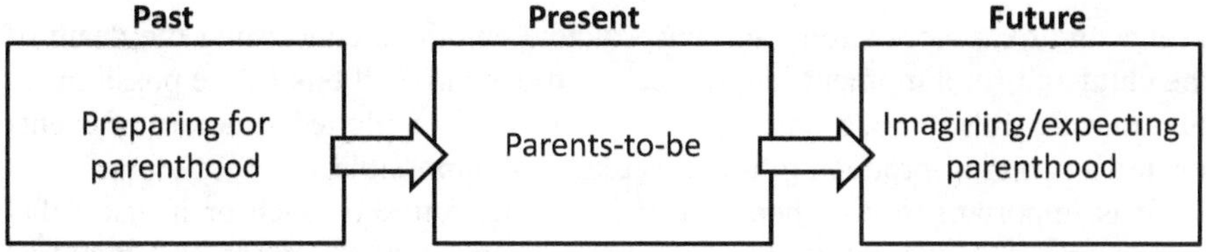

Figure 5.1 The "parents-to-be" as a signifying structure (cf. Romano 2009) before the death of the child; past, present and future are organized in a narrative plot towards the event of "becoming parents"

As "parents-to-be" the parents have created a common history of themselves wanting a child and preparing themselves for having one, while the future equals actual "parenthood," the wonderful new time to come where the couple becomes a family of three (or more). The present in this phase, described in Model 1, is a preparation stage where the parents-to-be buy the pram, changing table, baby clothes and toys, and decorate the nursery and perform all kinds of material practices in order to prepare for the child's arrival and the future to be. It is the child who will, after being born and now living in this world, qualify the performances and materiality.

What happens when the child dies—whether as a stillborn or very early in life—is that this whole future disappears and the narrative plotlines of the past that have been produced in order to point toward the welcoming of the child are rendered meaningless. All the roots pointing to this child as the only natural succession of all the parents' efforts are violently cut off rendering the preparation—however successful—pointless. Therefore, not only does the child disappear and with it the future as a prosperous time-space for the family-that-was-meant-to-be, but also the very meaning and signifying structures in life toward the future, as pointed to by Romano (2009). A child is a huge portion of potentiality, of potential reality; when the child dies potentiality itself vanishes and nothing can be fulfilled as foreseen. Instead a certain time-space (cf. Bakhtin 1981) is created solely containing the

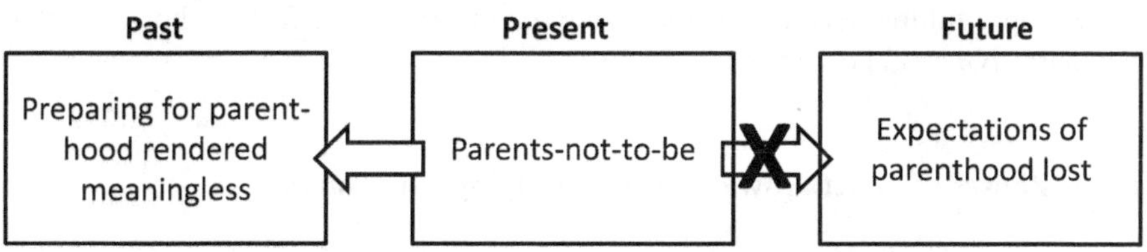

Figure 5.2 Just when the child has died this is the only focus of the parents. Past, present and future as a signifying time-space structure break down and with them the plot of "the parents-to-be" as a meaning-producing narrative

events and extremely strong and unspeakable emotions concerning the death of the child and for a moment (days, weeks—it depends) all talk of the possibilities of a meaningful life, of a narrative in which there is a plotted structure of events toward a meaning-producing future, is rendered impossible.

It is important to note here that it is not the future as such or in itself that disappears, it is the future in THIS version and contextualized and qualified through the production of these particular narratives. It is the future as a meaning-producing development of the plot inherent in the "becoming parents" narrative that vanishes. This is not a chronological structure, even if it seems that way. This is narrative time, particularly if the child is stillborn, but also if it dies at a very young age, meaning disappears in all three temporalities, not only in the future. This is a very crucial point. The future does not disappear but it loses the meaning ascribed to it by the parents: not only have they lost the child but they have lost themselves as signifying entities. They have transformed from "parents-to-be" to "parents who did not come into being" despite all the preparations, which now seem meaningless too.

Romano (2009), who was introduced in the beginning of this chapter, has some very strong reflections on death and bereavement. He makes the crucial point that in bereavement, where the cataclysm of death is felt by a human being, it is not only because we have lost *the other* (and to Romano we are only humans in so far as we are ourselves to "another"); rather, it is because when the other dies we are no longer needed by him or her and in this process we die not only to the deceased but to ourselves (Romano 2009: 115). So, losing the child is losing the constitutive role as a parent and thus the parents die.

We suggest that performing the rituals at Mindet.dk is to a great extent about reclaiming yourself as a parent. When a mother at Mindet.dk refers to herself as, for example, "Camilla, mother of Lucas" this is more than anything about claiming the right to be a mother even though your child has died. The collapsing of her role as the mother of Lucas, because Lucas died, leads to a repeated narrative (cf. Ricoeur 1980) where Camilla over and over again performs her motherhood by claiming the right to motherhood and performing herself as a (good) mother, for instance by "tucking in" her child at night-time through lighting a candle at Mindet.dk to tell the child "goodnight." This is one way of meeting the challenge of the empty future, performing and trying out in a ritualized fashion how to create new narratives and how to inhabit them.

Three Phases of Grief: Toward New Signifying Structures

We suggest that three different chronological phases of grief can be identified at Mindet.dk and we use these phases as focal points to make our analytical points. The phases are:

1. establishing a memory profile at Mindet.dk;
2. staying alive and performing parenthood;
3. developing new narratives/new directions in life.

Most of the memory profiles for the stillborn and dead children are established within a few days after death. While the parents, at this point, seem to be almost overwhelmed with grief, two other basic characteristics are discernible: first, life as the parents expected it to be has been turned upside down and, second, the parents are preoccupied with trying to understand what happened. In the first phase, "establishing a memory profile at Mindet.dk," these basics are dominant.

Speaking in terms of narrative, the parents have a narrative ready while expecting the child. As pointed out above, this narrative consists of expectations, ideas and matrices of happy family life, and so on. With the unexpected death of the child this narrative has to be replaced. At first, this cannot happen. Replacement will occur gradually in the years to come as life will unfold along new and, at the moment, indiscernible lines without the expected child and eventually events will be turned into a narrative with a once again meaning-producing plot structure (cf. Ricoeur 1980).

However, it seems as if one of the first activities of the parents at Mindet.dk is establishing a provisional, retrospective and coherent narrative the aim of which is to encapsulate the traumatizing and paralyzing narrative (the death of the child) by (re-)creating its explicit and brutal ending. This is often done by producing very detailed accounts (containing pictures and detailed descriptions of the moment of death) that in terms of timework seem to be attempts to "stretch" time and also may be seen as a first step in the process of *inventing memories*. Archeologist Tim Flohr Sørensen (2011) suggests that what he calls the "durable biographies of the deceased … are forced to build on reformulations of the past in the form of imaginary narratives" (Flohr Sørensen 2011: 162). We do not take this to be true of the narratives produced at Mindet.dk, firstly because we question the implied durability of the narratives, since these are narratives in progress, as is the picture created of the deceased. Flohr Sørensen argues that the narratives freeze the child at the age he or she was when he or she died. We cannot see this in our material. Rather, it seems that the age of the child differs according to the situations in which it is represented. Secondly, and most importantly, the narratives at Mindet.dk —both the long ones in the first phase and the shorter ones accompanying the lit candles—are not simply reformulations of the past in the form of imaginary narratives. Rather, they are little invented memory bytes narrating prototypical, mundane life experiences and turning them into memory-like narratives.

The provisional narrative is produced in a variety of ways and often in combination, as a "textual" presentation accounting for the different stages in the events leading up to the inevitable death of the child. These narratives can be very extensive[15] including detailed accounts of the hour-to-hour development and

15 http://mindet.dk/minde/lukas-andersen, accessed May 13, 2012.

unfolding of events, detailed descriptions of the child's diagnosis and medical conditions, events during the birth, some of this even presented as if narrated by the dead child itself.[16] While some are critical about whether the health personnel handled the events as best they could,[17] others are more calm in their narration of the events plotting the narrative, almost as if the right plotting will reveal a proper reason "why"; an understandable, if not bearable, version of the events leading to death or something that will, as a starting point, move the parents from standing in the middle of events to having moved past it, chronologically speaking. The parents have been told by others that their child has died and now they are creating a structural framing within which they can say it themselves: "This is my baby, and my baby is dead."

Some texts are poetry, written by the parents or a relative or friend[18] while others are the lyrics from a pop song.[19] Where the extensive narratives are accounts of events, these poetic pieces seem more like attempts to capture a feeling, a sense of something precious. However the form and structure, what seems important in terms of doing time, is the way these narratives focus on the sequencing of events and on the timing of death and the related events. Part of this process is about coming to terms with the loss of the child and producing some kind of closure, but at the same time there also is a strong sense of documenting the child through giving it some kind of duration, however short it may be. As long as the narrative moves on, the child is within reach or—in the case of the child not being a stillborn but having lived for at while, the narrative might bear witness to people (relatives, health personnel and others) having been influenced by the strong will or attitude of this child, which may be exemplified by the memory profile of a child who died at the age of three after fighting a terminal disease and whose narrative is the story of "Emil Superman."[20]

Besides the textual performances at Mindet.dk we see the use of photographs or drawings of the dead child, his or her grave (often referred to as a garden, for example, "Freya's Garden"[21]), besides more material documentation like hair-cuttings, handprints and footprints.[22] It is very important to note that these material representations of the child are not only tragic and sad. The parents are also full of pride and love, even joy, that this beautiful baby was bestowed on them. Like all new parents, the parents and Mindet.dk need to share the joy over their newborn

16 http://mindet.dk/node/102959, accessed May 13, 2012.

17 http://mindet.dk/node/11329 and http://mindet.dk/node/16136, accessed May 13, 2012.

18 http://mindet.dk/node/174949, accessed May 13, 2012.

19 http://mindet.dk/node/112366, accessed May 13, 2012.

20 See http://mindet.dk/minde/emil-supermand, accessed May 13, 2012.

21 See http://mindet.dk/minde/freya-catharine-bailey/gallery/82695, accessed May 13, 2012.

22 See http://mindet.dk/minde/elina, accessed May 13, 2012

and receive compliments—this they will get in rich amount from other parents at Mindet.dk.[23]

Besides establishing the child as a real person that was actually here for a while, however briefly, the existence of Mindet.dk and the narrative and performative actions of the parents seem to have yet another aim. They serve as devices for establishing the parents as parents, even though there is no baby. This aspect is supported by the repeated ritualization of lighting a candle as analyzed in the second phase.

In conclusion, what characterizes this first phase is to a great extent what we will call the invention of memories. When it comes to stillborns and very young children there are not many memories to begin with, but memory-like narratives are produced at different levels. Firstly, the documentation that turns a very short life (if any) into a thoroughly narrated life (or moment in time) in order to change a narrative which you can't make sense of (the paralyzing "WHY") into a narrative that you can relate to and cope with. In this process is laid the foundation for a creative agency for further life and narratives of the future. The narrative is processual and relational. Secondly, the dead child is turned into a superhuman agent existing[24] in a parallel universe that can be approached in a dialogic way, and is used as a positive force (guardian angel, confessor).

Where the first phase, described above, follows a relatively short time span after the death of the child, the second phase, "staying alive and performing parenthood," goes on through a longer period of time. After the first big performative step into Mindet.dk the repeated ritualizations are initiated; ritualizations aiming at maintaining the child—not only as a person who was here briefly and left the parents devastated but rather as a person who is in fact here and can be kept in close relation with this life through the performance of certain rituals such as the lighting of candles and, at the same time, ritualizations in which the mother performs "mother of somebody" thereby negotiating her role as a mother. These ritualizations happen most importantly, as we see it, through the parents' lighting of candles for their own child and for other children and also through postings in the forum and writing in other parents' guestbooks and getting appreciative messages in their own guestbook. These seem to have a double social quality; firstly, the mothers enroll themselves in the sociality of Mindet.dk, befriending the mothers of other "angels," a term often used to refer to the dead children and secondly, they continuously perform motherhood, not only through their signing ("Camilla, mother of Lucas") but also through commenting on the other children in their guestbooks, thereby imitating the practices of any new mother receiving and giving compliments.

---

23 http://mindet.dk/minde/ida/gaestebog, accessed May 13, 2012.

24 It is professor of cultural history Anders Gustavsson who reminds us of the difference between "life" after death and "existence" after death in his study of online memorial sites in Norway and Sweeden (Gustavsson 2011).

There seem to be two dominant and interrelated strategies for performing a good mother at Mindet.dk: through the lighting of candles you both demonstrate to other mourners that you care about your child and at the same time you "reassure" your child again and again that she or he will never be forgotten as you engage in the narrative process of inventing new memory fragments. This is particularly strong in the case of stillborns where instead of remembering things that have happened as the time has passed and the child grown older, things that could have happened in his or her lifetime are narrated in little episodes, for instance: "Today we went fishing. Couldn't help thinking that a boy like you would have loved that."[25] These repeated practices—narrative "stretching time"—reflect both duration and frequency in the concept of time work (cf. Flaherty 2011): The more frequently they are performed the more the child is kept present. Seligman et al. point to the repetition by stating that "the endless work of ritual is necessary precisely because the ordered world of ritual is inevitably only temporary" (Seligman et al. 2008: 30). Especially in the beginning of the mourning period, most candles are lit at night as a ritual telling the child goodnight and tucking her or him into bed. But as with many others, this ritual too must be repeated because the order established through the lighting is not permanent. Suffering and missing the child almost immediately returns and calls for new rituals to be performed.

However, we do see progression in these ritualizations and we do see certain structures that seems inevitable in order for the parents to move on with their lives. While in the beginning, many texts accompanying the lighting of candles are very intimate, very intense and seem very authentic we find it to be a tendency that a certain routinization occurs over time. Some parents begin posting their goodnight greeting with the same text night after night, and this "mantralization" stresses that in this frequently repeated ritualization there is no semantic meaning in the ritual—it is performing the action itself that carries the significations.[26]

At the same time we find in this period some postings that are more objective, such as, "It is your father's birthday today – we had grandma over for coffee," where the child has been turned into some kind of online friend or even a kind of confessor or human diary.[27] Even though there is still a lot of pain under the surface when candles are lit on red letter days or at Christmas a certain calmness becomes dominant.

It is very clear to us that in this phase of grief, Mindet.dk is a more or less isolated chronotope calling for other reflections and practices than those associated with everyday life and world at large. Actually we will argue, by introducing a third model (below), that the more the parents succeed in developing and maintaining Mindet.dk as a kind of parallel universe the more effectively can Mindet.dk be used as a closed ritualized space that can bring you in contact with

25 http://mindet.dk/minde/antoni-bottcher-larsen/lys, accessed May 13, 2012. See candle lit June 2, 2011.

26 We thank Jesper Oestergaard for suggesting his concept "mantralization" to us.

27 http://mindet.dk/minde/antoni-bottcher-larsen/lys, accessed May 13, 2012.

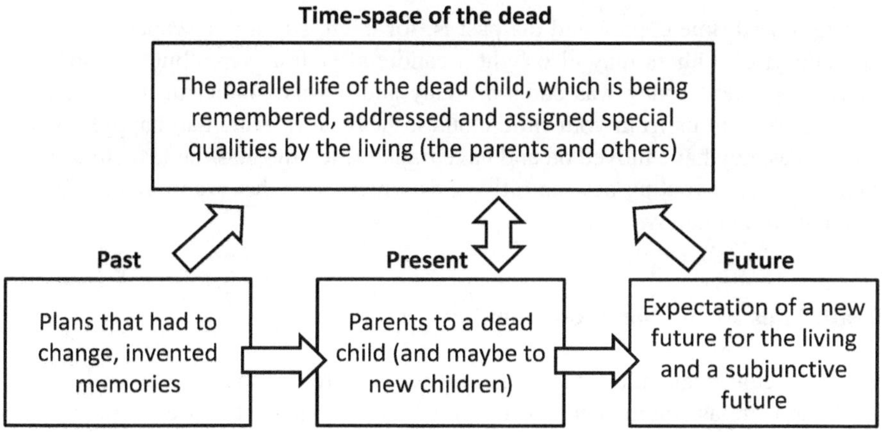

Figure 5.3 New signifying structures in the life of the parents in which the narrative consists of a plot containing a past once again is being reflected in meaningful ways in the present, creating possible directions into a set of meaning-making possibilities in the future

your dead child but also shut the child and the grief in so that it does not color all aspects of everyday life. Here a certain narrative is established in which the child is given a continued existence after death, not a consistent and chronological one but rather in an episodic manner through the invention of memories. The conception of a particular narrative, including a specific place for the dead child to continue existing, and the initiating of performative, ritualized practices in which a dialogue with the dead child is a possibility, is a common occurrence at online memorial sites in other countries as well, as demonstrated by Gustavsson's study of conceptions of faith as expressed on memorial websites in Norway and Sweden (Gustavsson 2011: 142f.), and the same maybe found in mourning practices found at children's graves (Flohr Sørensen 2011).

We can see that some of the parents seem very aware of how important these ways of keeping the child alive in a limited sense are, not only at Mindet.dk but also outside. One mother refers, in the forum, to herself having made a Christmas ornament (a glass bulb with the child's name and date of birth) to be hung on the tree symbolizing the dead child so that she can be part of the family Christmas in a low key but still in a central way and to whom the family "can sing while dancing around the tree."[28]

This development in the process of grieving and mourning at Mindet.dk may be observed in the following change in narrative. Those who light candles after four years and write something like: "Today was dad's birthday and you should have been here and the day was horrible because you weren't," will continue

28 http://mindet.dk/node/170560, accessed May 13, 2012.

having a hard time coping and the past is, for them, still an inescapable space of loss and grief. Others may also light a candle after four years but instead write something like: "Today was dad's birthday and it was wonderful that you were smiling down at us from your little cloud in heaven. It made dad happy!" which shows that they have moved on and have transformed the loss/the lost child into a creative force inventing new narratives, new memories that are more comforting than they are unhappy.

Death Ends a Life, Not a Relationhip[29]

Seligman et al. argue that ritual remains necessary to the ones that experience the world and life as fragmented and fractured and we suggest that even though the parents move on in life the fragmentation is permanent. "Ritual … is an endless work of creating a subjunctive world in overt tension with the world of lived experience" (Seligman et al. 2008: 28). In the first phase of ritualizations at Mindet.dk, the parents' timework consists of "stretching" time, imparting "duration" to the narrative of the child's short life, sorting out the "succession" of events regarding the child's life and death process and thereby negotiating "what happened" and, not least, appropriating death by being the ones to say: our baby died.

Romano reflects over this temporal dimension of separation:

> Bereavement is the absolute experience of separation: not a contingent and reversible separation, a remediable separation … but a separation that is in itself absolute, irreversible, irremediable, that happens in time and is sealed by time: the absolute of separation is temporal, because temporality is the only "dimension" in which an *absolute* separation is possible. In bereavement, such a separation is given us to *undergo*: the deceased has "left us", not to go "elsewhere" but because the time in which he was still present is henceforth "inaccessible" to us. (Romano 2009: 115)

In the second phase, the—often very frequently—repeated ritualizations of lighting a candle or other actions stress how Mindet.dk as a chronotope leaves a time-space for performing negotiations of new narratives and life practices, while still trying to find one's feet in the world and, not the least, negotiating the future position of the child. This is a matter of establishing a persistent and bearable relation to the dead child, and due to the fact that Mindet.dk as a chronotope is a social sphere (a community of mourners), part of this "establishing relations to/with the dead" is also about "establishing relations to/with the living" by way of supporting, comforting and sharing. At Mindet.dk a certain "angel"-terminology is dominant where the dead children are referred to as angels and narrated as located in the heavens and also supplied with certain superhuman qualities. One might

29 The former slogan at Mindet.dk.

even suggest that the ending of this phase—even though the borders to the third phase are fluid—is constituted by the child having been established in its afterlife existence, however nebulously.

The third phase, "New narratives/new directions in life" is, logically, a continuation of the second. Now, the narratives are in place, time is again becoming more differentiated and the child—and not least, the relationship to the child—can be maintained through less frequent, even though intense, ritualizations of the relationship. However, we cannot document the ultimate point of the third phase since the parents have moved on with their lives. They may still be active at Mindet.dk both in regard to their own lost child and in regard to showing respect and comfort to other grieving parents, but the performances, like lighting candles at bedtime or on red letter days, do not inform us of the new life they now live, though it may tell us that the dead child is not forgotten. The child has been given a parallel existence which the parents can and still do keep in touch with when they choose to do so—now and then, not very frequently, but they are continuously keeping hold (Walter 1997) even though they are moving on in their lives.

References

Bakhtin, Mikhail. 1981. Form of Time and Chronotope in the Novel, in M. Holquist (ed.) *The Dialogic Imagination: Four Essays*. Austin: University of Texas Press, 84–258.

Bateson, G. 2000. *Steps to an Ecology of Mind*. Chicago: University of Chicago Press.

Bell, C. 1997. *Ritual Perspectives and Dimensions*. New York: Oxford University Press.

Brooks, P. 1984. *Reading for the Plot*. Cambridge, MA: Harvard University Press.

Flaherty, M.G. 2011. *The Textures of Time: Agency and Temporal Experience*. Philadelphia: Temple University Press.

Flohr Sørensen, T. 2011. Sweet Dreams: Biographical Blanks and the Commemoration of Children. *Mortality* 16(2): 161–75.

Gustavsson, Anders. 2011. *Cultural Studies on Death and Dying in Scandinavia*. Oslo: Novus Press.

Heidegger, Martin. 1962. *Being and Time*. New York: Harper & Row.

Humphrey, C. and Laidlaw, J. 1994. *The Archetypal Actions of Ritual: A Theory of Ritual Illustrated by the Jain Rite of Worship*. New York: Oxford University Press.

Ingold, T. and Hallam, E. 2007. *Creativity and Cultural Improvisation*. Association of Social Anthropologists Monographs Series (44), London: Berg.

Ricoeur, P. 1975. *La métaphore vive*. Paris: Seuil.

Ricoeur, P. 1980. Narrative Time. *Critical Inquiry* 71: 169–90.

Ricoeur, P. 1984 [1983], 1985 [1984], 1988 [1985]. *Time and Narrative*, 3 vols, trans. Kathleen Blamey and David Pellauer. Chicago: University of Chicago Press.

Romano, C. 2009. *Event and World*. New York: Fordham University Press.

Seligman, A.B., Weller, R.P., Puett, M.J. and Simon, B. 2008. *Ritual and its Consequences: An Essay on the Limits of Sincerity*. New York: Oxford University Press.

Sjørslev, I. 2007. Ritual, Performance og Socialitet: En introduktion, in Sjørslev, I. (ed.) *Scener for Samvær: Ritualer, performance og socialitet*. Aarhus: Aarhus University Press, 9–34.

Storey, J. 2003. *Inventing Popular Culture: From Folklore to Globalization*. New York: Blackwell.

Walter, T. 1997. Letting Go and Keeping Hold: An Answer to Stroebe. *Mortality* 2(3): 263–6.

Chapter 6
Stones, Shamans and Pastors: Pagan and Baptist Temporalities of Death in Tribal India

Piers Vitebsky[1]

From Mute Stones to Speech of the Dead

From northwest Europe to southeast Asia and the Pacific, vertical stone monuments rise up as conspicuous traces of cultures whose material lives were otherwise highly perishable, and whose makers may have long disappeared. The materiality and inscrutability of stone offer a tantalising combination of permanency and silent witness, and it is common for archaeologists to write books with titles about mute stones finding voice (e.g. McKendrick 1983a, 1983b). This is animism as an academic and popular trope rather than as a serious cosmological understanding: in the world of scholarship, stones do not speak except through ancillary discourses supplied by indigenous texts or by archaeologists and other modern commentators themselves. Material objects do not so much convey messages, but rather appear to do so through projections from human interpreters.

However, in the cultures concerned, the speech of stones may be no mere trope and may present a serious philosophical or theological issue. The use of permanent and ephemeral material objects to bear such projections from those who create them may be an important clue to our understanding of how society structures time. In looking at changes in funeral practice among the Sora of tribal India, I shall distinguish two types of material presence: one is the memorial stone, the other the body of the shaman in trance, which partakes of what I prefer to call fleshliness or (borrowing from the language of theatre) physicality. Both of these are directly related to the immateriality of the dead whom they commemorate, and this relationship passes though the absence or presence, respectively, of dialogues between the living and the dead. The inert materiality of stones has limited scope when compared with the physicality of the shamans through whom the dead speak. Almost every day in every Sora village, a female shaman goes into trance

1 I have carried out research among the Sora since 1975 and during this period have been indebted to very many people who are listed elsewhere, mainly in Vitebsky (1993). The present chapter has been greatly inspired by recent conversations with Anastasia Piliavsky. I am grateful to the University of Tromsø for funding my most recent fieldwork.

and a succession of dead people speak through her mouth, using everyday speech to hold casual conversations or impassioned arguments with groups of the living who squat around her.

The Sora of Orissa number some 300,000 (which by Indian standards makes them a small minority group) and are part of a wider belt of so-called megalithic cultures of tribal India and southeast Asia (Fürer-Haimendorf 1943; Rousseleau 2012). They speak a language of the Austroasiatic family, which is related to Cambodian and to the languages of the Orang Asli of Malaya and of the hill tribes of Vietnam (Condominas 1994). Apart from the anomalous Khmer empire, the speakers of these languages consistently lie at the jungly margins of the great empires and mainstream civilisations of Asian history (cf. Scott 2009). In India, they are generally considered more ancient and more aboriginal than the much larger Aryan and Dravidian populations of the subcontinent. Though this picture is disputed for political as well as scholarly reasons, it is clear that their non-literate cultures lie well outside the Hindu mainstream.

But the secrets of Sora ethnohistory are in danger of becoming lost for ever. Since I started studying them in the 1970s, all young Sora in my study area have become Baptist Christians (while young Sora in nearby areas are joining neo-Hindu cults). The only remaining signs of the old culture of their grandparents (whom I shall call 'pagan' to distinguish them from both Christians and Hindus) are their large clusters of upright memorial stones, called *ganuar*. These are a residue of human actions, thoughts and emotions which would be entirely beyond recovery or imaginative reconstruction if it were not for some earlier accounts (mainly Elwin 1955) and my own experience of living with them for long periods during the 1970s, during which time I became a shaman's assistant. I give a detailed analysis of this religion in Vitebsky (1993). I have also made further short visits up to the present, and have witnessed how Baptist Christianity has consolidated its hold (Vitebsky 2008, 2011, in preparation). Thus though I use the ethnographic present, it should be remembered that the pagan culture I describe, and in which I have been so immersed, has become virtually extinct in the time I have known the area.

Sora memorial stones are unworked, rough and non-representational. One might expect that this would make them indefinitely open and potentially meaningful, as often happens in Hinduism, where an 'aniconic' stone may embody, indeed become, a god. But a Sora stone moves in the opposite direction, towards a diminution and eventual loss of meaning. Immediately upon death the body is cremated and the ashes buried. Apart from clothes and personal possessions, all material trace of the deceased is now concentrated into the memorial stone. On the day this stone is planted it is dressed in cloth, fed with liquor, rice and the blood of sacrificed buffalos, embraced, lamented, and raised upright as if representing the restoration of the deceased to some kind of life. But within weeks, despite recognisable differences in shape and size, people start to forget which stone is whose. The main message that emerges from the jumbled stack of stone slabs going back generations is their togetherness in the undifferentiated solidarity of the patrilineage.

Figure 6.1 An unattended *ganuar* between funerals: trying to identify old stones

Beyond this there is silence. But the future archaeologist would not realise that the reason the stones no longer have anything to say is that discourse about the deceased has migrated to another medium altogether, and one which will leave no material trace. In what may well be the most elaborate form of communication between the living and the dead ever documented anywhere in the world, the dead possess the bodies of shamans (mostly female) in trance and speak with the living at great length. This communication is necessary because the dead do not easily manage to stay with their lineage ancestors in the ideal way implied by the *ganuar* stones, but are pulled away by the circumstances of their individual deaths and the tangle of their personal relationships. The massive, unmoving statement of a lineage's clump of memorial stones is subverted by the frailty of the human body, the sociology of kinship, and the emotions of love and anger. The stones' presumption of eternity is overtaken by a quite different structuring of time.

Through dialogue, the dead regulate kinship, control inheritance and enforce morality. They do this in a particularly embodied way which is at the heart of the co-dependency between the dead and the living, by putting their soul-force into the grain which the living eat, but at the price of also causing the bodily pain of illness. All illnesses and deaths are caused by dead people who have themselves suffered from similar symptoms. Every time a person falls ill, it is a reminder of his attachment to a dead person who has started to eat his soul, thereby repeating the same symptoms in the patient from which he himself died (for convenience I use 'he' for patients and 'she' for shamans). The cure consists of offering him the soul of a sacrificial animal as a substitute for your own soul. If he accepts this substitute you recover; if not, you die from a repetition of his illness. Once you are dead, you will join him in whichever location on the landscape – tree, rock, stream – corresponds to the symptoms you now both share. Thus, people who die of smallpox, or are killed by a leopard, reside in the site of a Smallpox-Collective or Leopard-Collective, respectively. From there you will return as the most recent agent of that form of suffering to attack other living people with the same symptoms.

The dead themselves explain their own situation in graphic detail, as in this dialogue with a child who has died recently from leprosy (adapted from Vitebsky 1993: 171–2). Living and dead are involved in constructing a shared consciousness, with a close correlation between emotional distress in the mourner and a tendency to attack in the deceased. Hence in this dialogue it is assumed that the child will

Figure 6.2 A dead person speaks through the mouth of a shaman in trance, embraced by a tearful mourner

perpetuate her illness on to others. Her interlocutor tries to persuade her not to do this, but the girl's reply emphasises what everyone knows, that this is not a sign of a vicious person, but simply the existential property of a recently dead person.

> Dead Girl: [arriving from the Underworld, faintly] Mummy, where are my gold nose-rings?
>
> Living Aunt: They must have burned up in the pyre, darling, we looked but couldn't find them.
>
> Girl: [petulant tone] Why don't you show me my nose-rings?
>
> Aunt: They were so tiny. If I'd found them of course I'd show them to you. Oh my love, my darling, don't cause your own illness in others. Can you say that your mother and father didn't do enough sacrifices for you? They didn't turn their backs or refuse to help you, did they? Think of all those pigs they offered, all those chickens, goats, buffalos, my lovely child …
>
> Girl: [addressing her silent mother, and crying] Mummy, you were horrid to me, you scolded me, you called me Scar-Girl, you called me Leper-Girl, you said, 'You're a big girl now, why should I feed you when you sit around doing nothing?'
>
> Aunt: She didn't mean it, she couldn't help saying it. You were growing up and there were such a lot of chores to do.
>
> Girl: [sulky tone] I want my necklaces. [unreasonable childish tone] Why can't I have my nose-rings? I have to go digging, shovelling and levelling fields in the Underworld, all without my nose-rings. I came out in scars all over, my fingers started dropping off. That illness was passed on to me, that's how I got ill.
>
> Aunt: But don't you pass it on, don't you give it to your mother and little sisters!
>
> Girl: If I grab them I grab them, if I touch them I touch them, if I pass it on I pass it on: that's how it goes.
>
> Aunt: Your cough, your choking, your scars, your wounds, don't pass them on!
>
> Girl: [calling back as she returns to the Underworld] My Mummy doesn't care enough about me!

Each dead person has a double aspect: as an ancestor in a patrilineage, and as a member of a collective of fellow-victims of a cause of death. At first, the feeling of victimhood prevails in the dead person, who is aggressive and dangerous. Over several years the living mourners must persuade the deceased into a less aggrieved state of mind. At successive stages of the funeral, the shaman and her assistants sing and dance to 'redeem' (*tandi*) him from the anti-society of his fellow illness-victims, located somewhere on the landscape, and lead him into the company of his patrilineal ancestors in the Underworld. The assistants' chants (which I also used to sing) are performed in the persona of those ancestors. *Tandi* is the regular word used in financial transactions for redeeming a mortgage or a captive. Among various images of nurturing, sharing, and rescue by a war party, the ancestors will also use circular seeds shaped like coins to ransom the dead person back from his captors:

fathers	fellow-Ancestors
today	now
Sunday	Monday
Tuesday	Friday [i.e. every day]
our twin brother	our linked brother
our elder brother	our younger brother
our born child	our hatched child
by reaching	by merging
into our company	into our group
into our binding	into our bundle
let us comfort	let us escort
into our food hand-out	into our food share-out
all speak up	all speak out
we are right here	we are coming
the wealthy Headman	the prosperous Headman
circular-purse money	circular-purse cash
round-seed *tarab* money	round-seed *tarab* cash
let us brandish our swords	let us brandish our knives
let us speak like an advocate	let us speak for the defence
let us be forceful-mouthed	let us be forceful-toothed
let us lead him towards us	let us draw him by the arm towards us

This redemption is clearly an act of transformation. The deceased is being moved from one state to another, and in forceful language too. Yet the temporality of this act is complex, and the expressions meaning 'now' or 'today' signify not only the day in question but also a continual time, a perpetual present: now and always.

There are two senses to this temporality. One comes from the processual and partly cyclical nature of the Sora model of life and death. The emotional tone of dialogues between living and dead shifts over time. As both sides come to terms with their separation, the deceased becomes less aggressive and the mourner becomes less susceptible to his illness. My years of transcriptions show how living and dead gradually set each other at ease through a mutual persuasion, a process which has some striking parallels to a Freudian model of bereavement (Freud 1957, discussed in Vitebsky 1993: Chapter 10). This is not only an elaborate technique for structuring the processes of memory; it is equally a technique for regulating the pace of forgetting the individual while sustaining the social structure. An old ancestor who has been dead for many years no longer causes distress or illness. He continues to appear, but in order to support lineage identity. Finally he gives his name back to a new baby among his descendants (this is not a full reincarnation), becoming the child's protector against attacks from others – an agent of life rather than of death. Finally, a very old ancestor will die a second death in the Underworld, and turn into a butterfly. Butterflies are people who have been dead for so long that there is nobody left alive to remember them – they are memories without rememberers.

The other reason why the temporality of the dead person's transformation is complex lies in its tentative, provisional nature. Illnesses and deaths continue to occur because, despite the assertive language, each redemption is not fully successful at first. In the early years the dead person frequently 'reverts' (*gorod*) from the state of ancestorhood back to the form of his death-experience and continues to pass on his illness to others. These episodes represent swirls and eddies in the forward movement of time, and I believe they are linked to resurgences of grief among the living. It is only after years of dialogue that the deceased is content to perpetuate himself by passing on his name rather than his illness, and it is also at this irreversible stage of his redemption that the mourning of the living is completed.

By now the representation of the ancestor has moved very far from the unspeaking stone, not merely into the body of the shaman during her episodes of possession, but also into the living flesh of his descendants. The dead show a longing for corporeality, precisely because this is what they have lost. They exist in a state of deprivation and this longing, as acted out by the shaman, is often desperate and piteous. They crave to handle their old belongings such as axes and knives, to wear their old clothes and ornaments as well as newly bought cloths offered to them, and to drink alcohol and absorb the blood of sacrificed animals. Much dialogue is taken up with arguments about these offerings, with the dead complaining about the quality of things offered to them ('Your wine is sour, your animal is all skin and bones') and the living indignantly defending themselves ('Ai! Why are you complaining – didn't we give you a buffalo last month, and now you're making us ill again?').

Fleshliness and Embodiment

The attempt to assuage these desires passes through bodies and flesh – the other kind of 'materiality' which is so much more expressive and flexible than that of stone, yet leaves so much less long-term trace. It is through the pains of his body that the patient communes with the dead; it is the blood of the sacrificed animal which contains its soul and reaches the deceased; it is the shaman's body which absorbs the alcohol given to one ancestor after another during the seance, and which is drenched with tears by the mourners; it is the soul-force of fully-redeemed ancestors that enters the crops that nourish their descendants. When the shaman enters a trance, the voices which speak through her are taken as the authentic presence of the dead. But though the system is driven by dialogue, at moments of heightened emotion Sora mimesis of the dead can spill over beyond speech into bodily intimacy, and the shaman's envoicement becomes more like a full embodiment. These two levels of impersonation sometimes come together in desperate embraces between mourner and shaman, and even in amusing stories of naughty shamans who embody a dead lover in order to persuade an attractive mourner to sleep with them.

Thus expressions of kin bonds, of moral obligation and even of love pass via our bodily illness and suffering, which are taken as a reminder from a dead person of their claim to our attention. An ancestor whose nature is non-physical makes us ill by 'stroking' us. This quasi-physical contact causes the bodily reaction of an illness, and for the Sora this is the only explanation for illness and death. From an outside point of view this may appear as somatisation, a physicalisation of the psychic impact of a reminder of an emotional bond. But the terminology and logic of an encounter with a dead person short-circuits this roundabout psychological explanation by making the experience directly physical.

Perhaps one reason for the importance of perpetuating a dead person through speech is the absence in the old culture of photographs. There was no way of sustaining the presence of a dead person visually, for instance by hanging photos on the wall and making offerings to them as is done in local Hindu households and shops. Photographs (of a stilted studio kind) have more recently become popular among Sora just as the last shamanic rituals have been dying out. As part of this new photo-consciousness, young people have started requesting prints of my old photos of their forebears from the 1970s. In response, I have collaborated with Monosi Raika, an 80-year-old Sora friend, to produce an album of photos with captions in Sora (Vitebsky and Raika 2011). We initially produced 100 copies, but the reception has been so enthusiastic that we have been asked for many thousands more.

But even photographs suffer from the same fundamental limitation as a memorial stone: they can record but they cannot evolve, and like *ganuar* stones, my photos contain many old faces that are beyond identification. Pagan Sora practices of materiality and embodiment are based on evolution: they construct a time which is shaped into constantly repeated and finely-tuned circuits of evolution and resolution. Whatever the circumstances of your death, a way will be found to connect it causally to people who have died previously and to make it seem similar to their death. In this way, the specific, individual circumstances of each event are reduced to generic patterns. This is a technique of time-structuring which draws all contingent experience into patterns of repetition. Sora ritual is based on an explicit contrast between good and bad forms of repetition, and is designed to adjust the balance between these and progress from one to the other. When the dead perpetuate their illnesses onto us soon after their death, this is bad repetition; as they evolve to the point of giving their names to our babies, this is good repetition. Whereas some cosmologies seek to explain change and mutability, pagan Sora cosmology explains the constant form of events by pointing to the mechanisms which make those events repeat themselves in the lives of new persons. This interpretation requires flexibility to function, and dialogue is the technique that creates this flexibility and leads on to subsequent stages of this evolution. We can now understand butterflies as the final residue of the person, that which remains after all other experiences, suffering, personal possessions and other markers of biography have been passed on to others (Vitebsky 1993: Chapter 9).

As a tool of temporality, the *ganuar* stone has the most permanent, but also the least operational kind of materiality – even less than clothes and jewellery. The task of speaking is immediately transferred from this inert, hard medium to the soft physicality of the shaman's body. The stone is a starting point which leads to the flesh of the shaman; this fleshliness in turn provides the channel for dialogic speech. In the contrast of stone versus flesh, permanent versus perishable, it is the latter which has scope for narrative and development. Stones speak only of an unchanging, uneventful permanence, but shamans speak of the passage of time, the evolution of personal relations and the renewal of society. In a universe in which time is experienced and moulded through dialogue, time itself amounts to a way of talking about the process of decay and loss. In this sense, through their very permanence, stones do not participate in 'time' at all. Material things are alive, or temporal, only if there is decay, and conversely time can be present in material forms only to the extent that they are subject to this decay. This suggests that pagan Sora do not have a sense of time as a constant, given backdrop to human action and feeling. Rather, what we call time is experienced as fluctuations in the flesh, as they pass between life and decay, and between present intimacy and the final distance of the butterfly.

The Radical New Temporality of Baptist Christianity

A discourse of rupture is inherent in the rhetoric and experience of evangelical conversion (Robbins 2004). My own current quest, to be worked through in a forthcoming book (Vitebsky in preparation), is to understand why this new cosmology has taken over so rapidly and so totally from the previous cosmology which to my understanding had seemed so satisfying.

The takeover by Christianity in a new generation is directly linked to the arrival since the 1980s of roads, schooling, literacy, employment and a flood of development cash. When I first arrived there in the mid-1970s, pagan Sora were living under a system of governance which put them in a position of unremitting humiliation, ethnically, economically and politically. This power regime was starkly reflected in the shamans themselves. Even when operating their apparently un-Hindu tribal cosmology, Sora shamans gained their powers by marrying spirits in the Underworld who were high-caste Hindu rajas, policemen and government officers – that is, domesticated counterparts of the very people who oppressed the Sora above ground. Unlike the illiterate pagan Soras, these spirits could speak Oriya, the local administrative language, and write down names of miscreants in

Figure 6.3 Constructing a bigger and better village church

the logbooks of Underworld police stations. In other words, they used literacy as a magical technique of power and coercion. Thus, the standard Christian evangelical discourse of coming out of darkness into light maps well onto the wider Indian nation-building project and the Soras' own desire to emerge out of marginality into the mainstream (the fact that Baptist Christianity is an alternative to absorption into Hinduism is a further important aspect which cannot be discussed here). Older people are afraid to die because they know their children will not talk to them afterwards; but younger people feel liberated as they become literate in both Sora and Oriya and give up 'primitive' customs and subsistence agriculture to join the scramble for jobs as road labourers, Baptist pastors or government schoolteachers.

Baptist Christianity radically reconfigures pagan time. Pagans too have gods who made the world the way it is, but these gods then withdrew because there was nothing left for them to do. But Baptist time moves on from the Book of Genesis, not to a repeated or permanent present, but to a future end-time in which God and Jesus have a very active stake. This future time cannot be experienced but must nonetheless be striven towards through an agenda of social reconstruction and personal moral improvement. This discourse intentionally and aggressively cuts across the pagan principle of continuity through forms of repetition and recycling which merely lead back into the existing social structure. Funerals based on lineages are replaced by church events whose composition reflects the quite different structure of multiplying parishes. The aim of gatherings shifts from reconstituting the lineage after the rupture of death, to acting out the hierarchical, cell-like structure of the 180,000 living members of the Sora Baptist Association, encompassing hundreds of villages.

At the heart of this cosmological change lies an attack on the principles and logic of pagan causality. For Baptists, dialogues with the dead are forbidden because they reinforce attachment between living and dead, and thus encourage backsliding from the new regime. But they are also pointless, as there is little if anything left to discuss. The dead have no causal role in the fate of the living: illness is caused not by attachment and repetition, but by invisible 'insects' (germs, bacteria, viruses) in drinking water which can be divined only by non-Sora experts like doctors and health workers. Correspondingly, the living have no role in managing the ongoing state of the dead: death comes about through the 'will of God', who somehow avoids the challenge of dialogue and the blame that was previously heaped on the dead, and then takes care of the deceased in a way which is not well understood but which absolves the living of the obligation to keep enquiring about them.

The cosmology of radical change has clear political undertones, and early Sora converts were already composing hymns with a tone of liberation theology. When the government of newly independent India abolished an oppressive local form of bonded labour in the late 1940s, newly released Sora labourers started to circulate a song

nam sering nam sering	jijiang kambara maRilo bu
nam sering nam sering	red'red'ang kambara maRilo bu
nam sering nam sering	gîdrarung kambara maRilo bu, buñang
From today, from today	firewood-gathering slavery is finished
From today, from today	firewood-bundling slavery is finished
From today, from today	rice-handout slavery is finished, brother

Previously the shaman's assistants had used the word *nam*, 'today', to signify indefinite, perpetual time, and this is still what my colleagues and I were doing during funeral chants in the late 1970s. But by adding the postposition *sering* ('from') in this song, the released slaves were already turning 'today' into a benchmark term which leads away from a past towards a future: they took a usage from timeless ritual and used it to underpin a revolutionary new political time.

An early Sora Baptist pastor made a further adaptation of this song by using the resonance of political liberation to underpin decisive religious change. He composed an influential hymn which, after renouncing bad old practices such as blood sacrifice, drinking, smoking and fornicating, ends:

nam sering nam sering Kittung la	ompu ling mailenai
nam sering nam sering Kittung la	tegan tegaling la
nam sering nam sering Kittung la	maidan maidaling la
nam sering nam sering Kittung la	a bu'di bu'diling la
nam sering nam sering Kittung la	jeen jeeling la
nam sering nam sering Kittung la	anuyîp uyîpling la

From today, from today	I have joined God's group
From today, from today	I have joined God's team
From today, from today	I have joined God's [circle]
From today, from today	God's intelligence has given me intelligence [become my intelligence]
From today, from today	God's soul has given me soul
From today, from today	God's alertness has given me alertness

Here, the pastor builds on the era's mood of sudden and irreversible political liberation to give *nam* a spiritually repudiatory meaning. In the funeral chant, *nam* had set the temporal framework for a process of redemption (*tandi*) which was the ultimate goal of a pagan religion in which time itself seemed unchanging. In this new hymn, the loss of the funeral chant merges with the miseries of feudal exploitation. Both are resolved in tandem to lift the concept of redemption out of cyclical time and into a linear Christian time. The song about slavery said only what the liberation was *from*; the Christian adaptation says what the liberation would be *to*. Baptist Christianity would not only bring liberation from a cluster of oppressive entanglements (with needy ancestors as much as with exploitative outsiders), it would also complete the rupture by turning the Sora into new persons with a new kind of relationship to a new concept of divinity, all within a new time.

Just as our imagined archaeologist might expect, the shift to this cosmology finds an immediate reflection in the material marking of death. Where pagan cremation removes the body of the dead immediately and leaves nothing but the memorial stone, Christians are buried in tombs. These are elongated structures built up out of brick and cement, both recently introduced materials from the wider Indian world. They are arranged singly or in groups of loosely defined kin (including affines) beside paths, on waste ground near villages, or in the field of their owner. They are thus more individuated than the *ganuar* stones of the patrilineage, and their clustering does not represent any demarcated social group.

At a burial a pastor (always male) reads from the Bible and leads prayers. There is no drumming or dancing, and women are discouraged from weeping and singing old-style laments. After this the tomb will receive almost no attention ever again, apart from perhaps a clean-up and weeding on the anniversary of the death. Scraped into the cement of the tomb is the dead person's name with the dates of birth (often guesswork) and death. This most basic identity-card information is not developed in any discursive way – particularly by the dead themselves, who have no voice at all. Though pastors will make a lengthy speech at the drop of a hat, they do this in the genres of sermons, weddings and church rallies. There is no graveside oration about the dead person's life and works or public discussion of the social and emotional implications of the death. Baptist teaching directly inverts the logic as well as the emotionality of dialogues with the dead. Whereas prolonging this interaction brings solace to pagans, Baptists feel better by severing all contact. 'Remembering makes you ill', a young Christian told me simply, 'the Bible says forget and rejoice'.

The Christian tomb introduces a new form of materiality and temporality, both based on a new disengagement from the dead. At first sight, the neglect of the tomb might seem similar to that of a *ganuar* stone. The Baptist tomb also starts with a materiality which implies permanence. But this is not the pagan balance between the functions of a solid stone and a fleshly shaman. With its rejection of discursive commentary, the abandonment of the tomb points in a very different direction. The *ganuar* stone is silent because the voice of the deceased has moved on to the shaman in order to debate his ontology; the Christian tomb is silent in a way that leaves that ontology closed to enquiry.

Ordinary Baptists have to live with this muting of the dead, but sometimes have difficulty in letting go of the body. When it is laid in the tomb, the body is covered with money and with clothes which have been newly bought. This much is authorised by the pastors, but mourners may go further, against the pastor's instructions, and bundle up the dead person's old belongings (sandals, clothes, comb) to bury these too. Here again we see the Baptist rationale of avoiding sadness by disengagement rather than by engagement: 'He touched it, we may see it and feel sad.' But there is also a more menacing rationale: 'He may come back looking for it.' This shows a contrary continuation of pagan ideas of attachment, but no longer with a framework for dealing with it. Significantly, the personal belongings are buried, not in the actual grave but in a separate hole alongside. I interpret this as a separation between the offering of new clothes which can safely go into the grave because they have no history, and the used personal effects which are also a bundle of the dead person's attachments. These personal effects are the same kind of material props which helped to sustain communion with the dead among pagans, as the living continued to use them in daily life and showed them on demand whenever that dead person turned up in a trance.

So much for stones and tombs, both of which show the limitation of what can be gleaned from the hard material remains which await the future archaeologist. What of the fleshliness or physicality of the shaman? Though there is no space to develop an important argument which could be made here about gender, it is noticeable that the shaman is generally female, while the pastor is always male, and that this change is associated with an increase in hierarchy and authoritarianism. Whereas she wears ordinary clothes and works like an ordinary person, he arrives either by motorcycle or walking under the status symbol of a black umbrella, clutching a bible and prayerbook as a badge of office. He performs clothed like no one else, in freshly laundered and ironed white cotton like an Indian politician. The pastor's sermons and exhortations are perhaps no less performative than the trances of the shaman, but his performance is less fleshly (and to my eyes, also less emotionally intense). Where mourners may sometimes embrace the entranced shaman and cling to her howling with grief, the pastor remains separated behind a table, often on a raised platform. He does not embody those we have loved in his own body, and in his speech he does not enverb them.

Instead, he is a conduit for the words of God and Jesus, from whom he however always remains separate. Though people may have difficulty understanding passages from the Bible, these are not negotiated texts and there is no tradition of written commentary on them. The men on the platform say everything discursively and are the only source of authoritative interpretation of their own discourse. In effect, this is a narrowly controlled oral tradition grafted on to a fetishised written core. The old religion had fetishised writing from a position of weakness by making it the key technique of spiritual power in which literate helper-spirits enabled the articulacy of dialogue. Now, even while the universal literacy of young Christians makes the old-timers' awe at the mystique of writing seem pathetic, the inscription on the Baptist tomb takes the fetishisation of writing in a new direction, that of bureaucratic annotation.

Perhaps this was what the shamans' familiars used to do in their Underworld police stations: the pagans did not really understand what was being written except that it gave the writers power over those who were being written about. But the Christians have captured this power more directly, by mastering it themselves rather than relying on shamans' familiars to do it for them. Now that literacy is in the hands of living Sora, it serves a range of new techniques of precision. The programme of biblical texts for readings and sermons is printed in advance for every Sunday of the year at the Baptist Bungalow headquarters. As with Anderson's imagined community (1991), the same readings are being repeated on the same day in every church throughout Sora space. And in each village church the written agendas of meetings specify the timing of each speech and prayer, not even to the nearest five minutes, but to the nearest one minute.

Schieffelin (2002) analyses a similar attention to the micro-management of time in the present among evangelical Christians in Paua New Guinea, and convincingly explains this by a new connection between present and future. Whereas Sora pagan time has no end point or destination, Baptist Christianity introduces a new concept of end-time, which instead of the pagans' indefinite, partly cyclical present is future-oriented and teleological. Time is no longer structured through the pace of remembering and forgetting the dead. Instead of ancestral names, Christian babies receive names from the Bible which take no cues from family history but instead serve as moral templates. The Christian acquires none of the identity markers or the cumulative involvement of illness as constant reminders from ancestors which constitute so much of a pagan's biography and personhood. This new kind of person starts as a *tabula rasa* and develops a cumulative persona through Christian relationships and events.

What kind of timework does Baptist funeral practice facilitate? The Baptist relationship between present and future seems to have two contradictory paces, or to cluster round two opposite focal points. It is at the same time both instantaneous in the farewell at the burial, and indefinitely postponed in the expectation of the second coming. Between the day of the funeral and this remote and incomprehensible point in a future, Baptist funeral practice omits much of the middle ground of living memory. This is precisely where the pagan religion put all the emphasis, with its elaborate discursive procedures of dialogue and negotiation.

Pagan time is a time of lineage continuity and replacement, but, as the earlier mention of Freud suggests, it is also the time of bereavement therapy. Its measured pace is linked to the dynamics of living memory and to a specific way of moulding a long drawn-out process of forgetting. The Baptist rejection of dialogues with the dead entails a rejection of the pagan structure of time itself and its replacement by a new structure of time to match this reduced engagement. Since formally there is nothing – no structure, no phases, no acknowledged experience – between the moment of burial and the second coming, time itself becomes more abstract and less amenable to understanding. Even pastors seem uncertain about when the dead finally reach Jesus, or how long they hang around while waiting. However hard I press them, I find Baptists' discourses about cosmic time weak and hesitant, I think because of the thinness of the theory which supports them. This is perhaps also a sign of a new religion which has won demographic supremacy faster than it has been able to consolidate the relationship between doctrine and experience.

Baptists have no use for unmarked, aniconic stones, since they have no shamans to draw out those stones' latent meaning. The first time I saw *ganuar* stones dismantled by militant Baptists in the 1990s and re-used for cattle fencing, I experienced a shock of blasphemy. Looking back, I understand that these stones were being removed from their pagan relationship of contrast with the fleshly time of shamanic embraces, and were being forced to take up a new role in Christian time. But though I could not control my feelings at that moment, perhaps I was wrong to see this as blasphemy rather than as a sign that the stones had merely become irrelevant. Young Sora who have been brought up tightly within the church now know nothing at all of the old religion, and assure me that their parents used to 'walk in darkness' and 'worship devils'. There is room in Baptist cosmology for a positive past, but this is only the past of ancient Israel in which one accepts the doings of ancient Jewish priests. The retrospective photograph album which

Monosi Raika and I produced (2011) also contains extensive texts of shamanic rituals, thus offering the only alternative reading matter to the Bible in the Sora language. Though he was a pioneer Baptist convert and church leader in his youth, Monosi now laments the loss of the earlier culture which he helped to undermine (see Vitebsky 2011) and is using my old field notes to help him 'preserve' it. Monosi is the most theologically insightful Sora alive, and through talking to me he has come to his own conclusion, that pagan Sora religion is similar and equal to Old Testament Judaism, as the stage reached by each religion before the coming of Jesus.

Monosi is one of the Baptists old enough to have been formed under paganism, and for such people there are tensions which are hard to process. It is impossible to know how the dead are faring, because there is no channel for them to come and tell us. The only way left for them to reclaim something of the voice that was silenced by the Christian funeral format is to appear in dreams. In an article about these tensions (Vitebsky 2008), I explored an episode which revealed the sadness of a young Baptist called Paranto, who had been prevented from mourning his father in a pagan way. On another occasion, Paranto told me about a visit from his father in a dream:

> It was as if we were meeting in waking life:
> – 'You've died, where have you come from?'
> – 'I'm not dead,' he said, 'I remain, I exist (*ñen dakutenai, ñen daku*)'.
> – I got up and looked around: it was the middle of the night – (amazed tone) there was nobody there! I cried, I was very sad: 'I thought you were dead!' I met him under the tree on the way to his favourite drinking place: 'Ai! where are you going?'
> – 'I'm just wandering around'. He looked just as he had when he was healthy.
> – 'But how come your body was so sick, and now you're healthy again?'
> – 'It's alright, I'm fine now'.
> I looked around, there was nobody.
> [Piers asks: Were you frightened?] I was, but he didn't scold me and didn't harm me. We just talked, the two of us, and sometimes we cried together.

The dream narrative continues for some time more. The son receives a reassurance he desperately craves, and which Baptist Christianity is unable to give. But this reassurance can come about only within an imitation of the processual pagan time-frame of stages of transformation. As the narrative progresses Paranto will learn, not just that his father has been healed of his illness, but also that he now resides in a pleasant place – not in the sky with Jesus as Paranto is supposed to believe in waking life, but in the Underworld with his ancestors who led him there, as if he had undergone all the stages of the pagan rite of redemption.

Even in its new radical Christian guise, the Sora evidence still directs our concern with temporality towards the presence or absence of ideas of transition between phases and of techniques for enabling such transitions, and thus towards the shape of the time within which such transitions take place. I believe that this can also guide our understanding of the relationship between time and materiality. In the current Sora shift from memorial stone to Bible and from shaman's embraceable naked torso to distant, white-costumed pastor, what emerges as important is not so much materiality (or even physicality) itself, since this now appears as an index of something much deeper. Rather, it is about degrees of an ability to bear what we might call the projection and evolution of human relationships. The structure of time, whether pagan or Christian, arises through the onward development of such projections. In both systems of metaphysics, time must serve as the ground for ideas of resolution. Pagans and Christians agree in principle that life contains a suffering which must be assuaged after the event by a process of hard work which they both call by exactly the same term, redemption. For Christians just as for pagans, this is salvation in a spiritual sense, conceived as a metaphorical extension of the literal buying back of a person from captivity (e.g. Exodus 21:8). Pagans and Baptists disagree about the causality of that suffering and about the source of the authority and charisma which can redeem it. But even more encompassingly, they disagree about the very concept of time within which this redemption must take place. For pagans the entire cosmos, indeed existence itself, is ancestorised: it is manifested and functions through dialogues with the dead. The time of redemption takes off from the present and happens within the span of a group of closely interlocking lifetimes and by means of negotiation between those persons. For Christians, redemption takes place in a time which is discontinuous with the present, in a transcendent domain which lies beyond both experience and society.

References

Anderson, B. 1991. *Imagined Communities: Reflections on the Origin and Spread of Nationalism*. London: Verso (revised edition).

Condominas, G. 1994. *We Have Eaten the Forest: The Story of a Montagnard Village in the Central Highlands of Vietnam*. New York: Kodansha Amer Inc (original French publication 1975).

Elwin, V. 1955. *The Religion of an Indian Tribe*. Bombay: Oxford University Press.

Freud, S. 1957. Mourning and melancholia, in *The Standard Edition of the Complete Psychological Works of Sigmund Freud*, translated by J. Strachey. London: Hogarth Press, vol. 14, 239–58 (original German publication 1917).

Fürer-Haimendorf, C. von. 1943. Megalithic ritual among the Gadabas and Bondos of Orissa. *Journal and Proceedings of the Royal Asiatic Society of Bengal*, 9, 149–78.

McKendrick, P. 1983a. *The Greek Stones Speak: The Story of Archaeology in Greek Lands*. New York: W.W. Norton.

McKendrick, P. 1983b. *The Mute Stones Speak: The Story of Archaeology in Italy*. New York: W.W. Norton.

Robbins, J. 2004. *Becoming Sinners: Christianity and Moral Torment in a Papua New Guinea Society*. Berkeley: University of California Press.

Rousseleau, R. 2012. Megalithic landscapes, cultures and identity in northeast India, in *Nature, Environment and Society: Conservation, Governance and Transformation in India*, edited by N. Lainé and T.B. Subba. Delhi: Orient Blackswan, 17–35 .

Schieffelin, B. 2002. Marking time: the dichotomizing discourse of multiple temporalities. *Current Anthropology*, 43 Supplement, S5–17.

Scott, J.C. 2009. *The Art of Not Being Governed: An Anarchist History of Upland Southeast Asia*. New Haven: Yale University Press.

Vitebsky, P. 1993. *Dialogues with the Dead: The Discussion of Mortality among the Sora of Eastern India*. Cambridge: Cambridge University Press; Delhi: Foundation Press.

Vitebsky, P. 2008. Loving and forgetting: moments of inarticulacy in tribal India. *Journal of the Royal Anthropological Institute*, 14, 243–61.

Vitebsky, P. 2011. Repeated returns and special friends: from mythic encounter to shared history, in *Returns to the Field: Multitemporal Research and Contemporary Anthropology*, edited by S. Howell and A. Talle. Bloomington: Indiana University Press, 180–202.

Vitebsky, P. In preparation. *Living without the Dead: Changing Forms of Love, Loss and Redemption in Tribal India*. University of Chicago Press (provisional title).

Vitebsky, P. and M. Raika. 2011. *Jujunji do yuyunji a banuddin: Sora jattin a sanskruti (Sora beran batte, aboi tanub)* [Indigenous Knowledge: A Handbook of Sora Culture (in Sora, Part I)]. Visakhapatnam: privately printed.

Chapter 7
Dealing with Dead Saints

Mikkel Bille

Introduction

The grave of Sabah al-Sayydiiyyn near Bir Hamad, south of the world heritage site of Petra in southern Jordan illustrates a dilemma common to the neighbouring Muslim communities: how to deal with dead saints in a proper Islamic way? As the oral tradition goes, Sabah lived in the early half of the nineteenth century. Whilst alive, Sabah was a good person, helping others and taking care of the orphans in the area. Yet, he was by no means a person who was likely to live on in local history centuries later. When he died, he was buried with a heap of stones piled on top of his grave, like others at the Hamad family graveyard near Bir Hamad. Immediately after the burial, however, water flowed from the ground in this otherwise barren area. People rejoiced, and washed and drank from the water. This was a clear sign from God that Sabah had indeed been bestowed with God's *Baraka*, God's blessing. This *Baraka* would now be materialized in the water flowing from the dead man's grave. Even in death, the dead person would emit this somewhat mystical blessing from God, in various ways used to describe a particular divine energy or presence (Brown 1981: 86; Geertz 1968: 33, 44; Makris 2006: 145–9; Roberts and Roberts 2003: 24–8; Stewart 2001: 236). However, this does not mean that a visitor at the grave *will* be healed for sure, simply by drinking the water, only that the *potential* for healing has radically increased.

Some time after this event, one man decided to close the spring again with an upright stone, visible on the right side of Figure 7.1. This, however, caused the man to become barren for the rest of his life. To the residents in the area, this was a sign that disrespecting the dead would be met with immediate retribution. The events at the grave, both fortunate and unfortunate, meant that Sabah came to be considered among the 'friends of God', in Arabic, *awliya* (singular *weli*). Since then, people in need have travelled to the saint grave to seek divine blessing by praying and offering gifts. I here translate the term *weli* into 'saint' for convenience, although it covers a variety of religious figures in the Islamic tradition, such as Sufi, dervish and fakir. This translation into 'saint' is commonly used by scholars and is not related to a Christian theological reading (Bandak and Bille in press; Denny 1988; Makris 2006: 143; Meri 2002; Stauth 2004).

In recent decades, however, a more scriptural understanding of Islam has spread through mosques, satellite television and schools in the area, often termed the Islamic revival (cf. Mahmood 2005; Hirschkind 2006). For an increasing

Figure 7.1 Saint's grave in Bir Hamad, Jordan (photograph by author)

number of people, saintly intermediation is considered contrary to Islamic teachings. The social life of dealing with the dead is here shaped in terms of the moral economy of categorizing time; saint intermediation belonging to 'the age of ignorance', *jahiliyyah*, rather than the age of Islamic knowledge. Dealing with the dead saints thus ties in with the emerging moral evaluation of the propriety, even piety, of doing so.

While some people dismiss the actions at the graves as heretic, without hesitation, others are caught in a paradox of also finding a source of personal identity in the stories of saintly intermediations, since the stories may concern family members, or the saint may be a forefather. The issue for those people who no longer go to saint graves or never did, as well as the few that still do, is that they are forced to deal with the immediacy of the graves: the presence of elaborate material ornamentation, and the stories of events that have taken place. The very matter of death, so to speak, as well as the stories of potent blessing, force people to take a stance on the materiality of death (Hertz 1960 [1907]: 27; Hallam and Hockey 2001; Hockey et al. 2010). In this case, it is about taking a stance on the excess that spills over into the lives of the living with the potential actions of the dead saint. This leads to question in what sense the dead and the material anchors at the grave produce 'agency', and how to deal with it.

In this chapter I discuss saint mediation and the stories of saintly intervention in people's lives in southern Jordan, as they relate to questions of agency, and thereby following categorizations of time into eras of ignorance and of Islamic knowledge.[1] Exploring the question of the so-called agency of things will point to the intricate ways in which objects, practices and categorizations of time are ambivalently interwoven in the social processes of dealing with death. Rather than engaging with the corporeality of the dead body as such, or conceptions of death in Islamic theology, I will instead argue that a moral evaluation inherent in categorizations of time and the potentiality of blessing play crucial roles in shaping regimes of knowledge and pious subjects. In essence, I will be relating the deeds of dead saints to categorizations of time by exploring, ethnographically, some assumptions about the term 'agency'.

The Agency of Things

Let us start with the term agency. Over the last two decades or so, debates in various branches of material culture studies have discussed whether or not things have agency and, if so, on what premises (e.g. Gell 1992, 1998; Knappett and Malafouris 2008; Latour 1992). Most often these discussions are tied to the question of the role of intentionality. In one dominant perspective, agency presumes intentionality. Here, things do not have agency, as things cannot consciously *choose* to do otherwise; choice or will belongs to the faculty of human/animal cognition. 'Agency' thus becomes the capacity to intentionally initiate actions (Johannsen 2012). Oddly enough, this perspective is rarely confronted with a dictionary definition of the term from, for example, the *Oxford English Dictionary* (2010):

> Agency: 2. [mass noun] action or intervention producing a particular effect: Canals carved by the agency of running water. [count noun] a thing or person that acts to produce a particular result: The movies could be an agency moulding the values of the public.

Naturally, a dictionary definition is not written in stone, yet it should lead to some sort of reflection on the tenability of the position that intentionality is a prerequisite for agency. Water cannot be said to carve the canals intentionally.

In recent decades, the social sciences have also increasingly shifted away from seeing agency as 'capacity' and to focus instead on 'causation'. From this perspective, things are continuously 'enacted' and have (or produce) agency as agents or actors (Gell 1998; Laet and Mol 2000; Law and Mol 2008; Latour 2005). An agent in this perspective is – in accordance with the dictionary entry – what causes an event to happen where 'an entity counts as an actor if it makes a perceptible difference' (Law and Mol 2008: 58).

1 The material for this chapter was gathered during my one-year doctoral fieldwork in 2006–7 and, again, during two months of fieldwork in spring 2011 in the Sharah mountains from Shobak to Thoghrah.

Yet intentionality is still a dominant concept in the discussions. For instance, Alfred Gell (1998: 16–23) makes a distinction between primary and secondary agents: a primary agent has the intention of affecting its surroundings, while a secondary agent is one that causes an effect without implying any intent, even though this agent may mediate the intention of its maker. Other ways of subdividing categories of agency have been promoted by talking about 'conscious agency' and 'effective agency' (Robb 2004: 132), while still actively dealing with intentionality. As Lambros Malafouris critically notes, most often 'intention in action' does not precede action or matter, but unfolds *with* action (Malafouris 2008: 31–5; Johannsen 2012: 313). Nonetheless, in this perspective things as such do not have intentionality, even if they may produce agency.

From a third perspective, things do have intentionality, but of a different kind – a technological rather than a human intentionality. With inspiration from phenomenology, the argument is that intentionality (at least etymologically) simply means *to direct* (Verbeek 2008: 95; Ihde 1990). As a technological object rainwater canals has intentionality insofar as it directs water to other places.

When it comes to death, however, there is something bewildering about these discussions. On the one hand, there is a commonly held notion that the dead cannot speak or act, which is in line with the initial perspective above. Yet the dead body does, nevertheless, alter its material composition, by decomposing, smelling or attracting animals, in a way that also forces people to deal with it as both a moral obligation and bacterial phenomenon (Hertz 1960 [1907]: 27). People tend to the dead, preparing them for the last rites, carrying them to their grave, and so on. Several scholars thus argue that the dead body acts within its social context, in line with the second and third perspective above (Hallam and Hockey 2001; Harper 2010; Hockey et al. 2010; Sørensen 2009).

Yet, while these discussions may have nuanced our understanding of intentionality and agency as pertaining to different realms of humanity and materiality, or even merged them, we are still left with the question of how to deal with non-mechanical responses, such as becoming barren, initiated by engagements with the non-living.

The Potential for Agency

Indeed, as the story of the events at Sabah's grave goes, and according to many other similar ones, the dead and their graves can act, even in impulsive response, to actions oriented towards their material anchors among the living. The response may not be causal or even normally possible, and the dead may not act alone, but in relation to other human and non-human agents linked up with them. God's will is channelled through the dead saint to the material anchors where the devotee's performances and material offerings may connect them to the saint. While digging a grave can logically cause water to flow from arid ground, barrenness does not seem to be the direct mechanical response to stopping the water from flowing. The question of 'agency' becomes potently unclear: was it God, the saint, Satan, or pure (mis)fortune? *Who did it?*

In this Islamic context, these post-death events are often taken as a sign of the dead person's closeness to God – in life and in death. People subsequently go to the grave to heal illnesses or change misfortunes by seeking *Baraka* (Goldziher 1967: 209–38, 1971: 255–341). In either collective or individual pilgrimages, people go there to offer their prayers to the deceased, light candles, burn incense, leave a coin or bead, tie a piece of white or green cloth on a wooden beam on the grave, and possibly slaughter an animal, to mediate between the Divine and earthly worlds. These practices have been extensively studied in anthropology in terms of 'folk Islam' as opposed to various sorts of 'scriptural', 'fundamentalist' or 'puritan' Islam (most famously by Gellner 1963; Gellner 1981; Geertz 1968; but see Makris 2006: 49–50).

As a friend of God, there is simply *too much* human agency for the person to cease at death. The good deeds and actions of the person continue after death. This excess of human agency encountered at the grave is neither intentional nor simply mechanical in the above perspectives. It is not a water canal that mechanically prevents water from overflowing the cities. But neither is there a living person who intentionally does something. A material analogy to this excess at the graves would be mechanical devices that should perform a simple task but which, for various reasons, only do so sporadically to an improbable extent beyond the realms of normality, comprehension, or even desire.

What I point at here is the place of potentiality in the discussion of agency. The performer of the intermediation is not certain to have his or her wishes come true. That is, from the perspective of the living devotee, the capacity of the saint's grave to impact on their lives is not understood in any mechanical sense. Its agency is not so much tied to questions of *intentionality* – what the dead saint chooses to do – as to the *potentiality* of what he or she *could* do through the *Baraka* at the grave as a connecting force between dead saint and the living supplicant; a hope for potent actuality. No one knows why it happens. It just does, on certain occasions. Emulation of practices that previously led to success is no guarantee so the question of whether one is doing too much or too little for intermediation is hence never fully answered (cf. Geary 1994: 95–124). It is, in this case, the grave, or shrine, as a material extension of the dead saint that works as a site for intermediating God's *Baraka*. But the question my informants ask is not so much why, but if, something will happen. Thus, the approach here is one that draws on the notion of agency as causality, while also emphasizing irregularity and non-causal relations between act and reaction at the grave, through the role of potentiality.

The potency of the saint grave, in essence, relies on more than simply passing the site and being affected by the *Baraka*. It is a temporal actuality in the sense that the grave does not always 'have agency' and fulfil people's needs. Instead, even if only on extremely rare occasions, the saint grave may 'eject' its intermediary potentiality onto the devotee. That is, in certain constellations of things and people the graves may indeed produce agency, but they do not emit this force all the time.

One can perhaps think of the saint grave not just as a fixed, coherent object, but as 'enacted' – to evoke Annemarie Mol's term (2002) – in an non-mechanical way as an 'eject'; an ejection that promotes temporality and potentiality as elements of understanding materiality.

In that sense, to the devotee, the saint grave is an intermediary that may influence divine powers that are beyond what the individual dares confront directly. This follows other parts of social life in the Middle East where the concept of *wasta* (generally understood as nepotism or connections) holds a central role in mediating social relations and power, and indeed was the term used by my informants to describe the nature of the mediation (Cunningham and Sarayrah 1993; Hammoudi 1997; Shenoda 2008; Meyer and Moors 2006).

Living with the Saints

Although the area around Petra is most famous for the Nabataean and Roman monuments, it is also known for its religious sites. In the mountainous landscape around Petra, there are currently around 30 saint graves, shrines and sites, such as sacred trees and rocks, which can be visited on collective or individual pilgrimages. While most of the saint graves are modestly built, such as Sabah's grave, there are a few solidly built shrines with a dome (*qubba*) for prominent historical figures. Most notable among these are those relating to biblical history, such as the shrine of Moses' brother, Aaron, on Jebel Haroun, the rock where Moses struck his stick to get water in Wadi Mousa, or, from Islamic history, the shrine of Abu Suleiman al Darani in Abu Makhtoob.[2]

These shrines used to be centres of pilgrimages. The largest and most famous pilgrimage was the biannual collective pilgrimage to the shrine of Aaron on Jebel Haroun. The pilgrimage to Jebel Haroun is so renowned that it was part of the successful application to include the oral and intangible heritage of the Petra Bedouin on UNESCO's intangible heritage list (Bille 2012). Twice a year, people would ignore gender separation or tribal feuds and meet in collective song and dance all the way through the mountains to the top of Jebel Haroun. Here, like at the many other graves and shrines, incense was burned, animals sacrificed and prayers offered that addressed the dead saint, calling to him to ask God for healing. The widespread practice of tying knots with cloth on the grave in this region is an example of what James Frazer called contagious magic which, unlike magic characterized by imitation, is based on the idea that material objects that have been in contact with a person may produce a potent bond with that person and act

2 It should be noted that many of the graves are often known by different names, stories or tribal affiliations depending on the informant; one informant, for example, called it the grave of Sheikh Ibrahim al-Khasheibi. This is perhaps an important characteristic as it enables different people to infer different meanings from the graves and thus sustain its relevance.

on his/her behalf even at a distance (Frazer 1993 [1922]: 11). Michael Gilsenan further writes that 'The pilgrim hopes to obtain a direct experience of *Baraka* as much through touching the tomb or cloth over the shrine … as by the hoped-for favour' (Gilsenan 2000: 605); that is, the hope is equally distributed in the faculty of the saint as in the sensation of matter. Why or when it happens is unknown, but the connection the saint grave's *Baraka* potentially offers is crucial. It is not just the saint acting; it is a connection offered through the grave.

Some of the first saint graves I was introduced to during my fieldwork on sainthood and settlement in the spring of 2011 were located in the centre of the city named Wadi Mousa.[3] At one of the most congested places between the upper and lower parts of the town was a graveyard with a small shrine for a saint called al-Haseni. Henna was smeared on the walls as a sign of past rituals, but the green cloth covering the grave that normally denotes saintly status was carelessly discarded on the ground following a grave robber's excavation. Clearly, the grave was no longer in use for intermediation (see also Canaan 1929: 207).

The second grave was barely visible under a tree called Shajarat Ataya, just 150 metres from al-Haseni's tomb.[4] The (*ephedra aphylla*) tree is so big that it conceals the grave. The fecundity of the tree acts as proof of the fact that the person was blessed.[5] While both sites also highlight the ability of the material world to tie inhabitants to their tribal history and geographical belonging, the first case emphasizes the current dilapidation of the graves, as intermediation practices have stopped and grave robbers have taken over. The other case shows the continued life of the dead through the construction of a causal relationship between the fertility of the soil and the *Baraka* bestowed on a person by God, although it is no longer used for intermediation.

Clifford Geertz (1968: 45) reminds us that there are two complementary ways of perceiving *Baraka*; one is the genealogical, whereby it is inherited from one charismatic person – most often the Prophet Mohammed – down the patrilineal line; the other is the miraculous, which is linked with specific events in people's lives. This double-sidedness ties a wide range of people to the graves. Some of the graves

3 The city is named after the valley it is located in; Wadi Mousa means Valley of Moses.

4 The same name is also given to another, now disappeared, tree in the village (Canaan 1929: 207; Al-Salameen and Falahat 2009: 190–191). Some places are known by different names depending on tribal association, such as the site al-Bawwat; according to my informants from the Sayydiiyyn tribe it is located in Taibeh and visited by the inhabitants, but my the informants in Taibeh had no knowledge of this. Most sources indicate that al-Bawwat is another name for al-Fuqara at Ayn Amuun (Canaan 1930: 179). Furthermore, other sites are presented in the literature on saints in the region, particular Tawfiik Canaan's pioneering work, which mentions the grave of Abu Hmeid in Beidha but which no one in the area has any knowledge of today, even the Hmeid branch of the Ammarin, among whom I lived for a year during my doctoral fieldwork.

5 According to oral traditions, there is a cave full of gold close to this, which will open when a man named Abdullah bin Hamad al-Naimi arrives.

were predominantly visited by members of the deceased's tribe, such as Salem Awath al-Ammarin's grave near Bir Madhkur, and act both as places for divine mediation and spatial anchors of tribal and family history. Other graves, such as those of al-Fuqara at Ayn Amuun, described below, are famous for being powerful enough to have attracted people from all over the country, even Saudi Arabia.

The saint can be seen as a person from whom the devotee wants something, even if the person granting the wish has the unusual feature of being dead (Maniura 2009: 646). The graves are, however, not only sites of benevolence, as the case of Sabah's grave attests. People who harm or destroy things at the grave of the saint are likewise prone to misfortune. The grave of the saint known as Ghannam in the area of Gerendal in Wadi Arabah is another powerful example of this.[6] As elsewhere, a wooden beam is fixed to the saint's grave from which a piece of cloth is to be hung, which materializes and creates the potential contact between living and dead. As the story goes, a man took this beam and used it in his tent. The next day, he was unable to move, paralysed, thus showing that 'to act is not to master, for the results of what is being done are often unexpected' (Law and Mol 2008: 58). He was then advised to revisit Ghannam's grave and pray. He returned the beam to the grave and sacrificed an animal that his companions could eat for dinner, after which he was cured.

This story works not only in a functional way to prevent further destruction of the shrines but also to prove the potential potency that the graves materialize. This is not simply a potential that is initiated in the complex practices of rituals that combines prayer, offering and 'contact magic', but equally one that may simply flow from the grave as a constant blessing that needs to be dealt with. The three saint graves in al-Mureibit cemetery also illustrate this point. They are the passing point of seasonal migration for some pastoral nomads, as they move between the valley and the plateau through the mountains during autumn and spring. People lead the animals through the graveyard to graze in the area, thereby physically internalizing the *Baraka*. One story goes that once there was a family who had not

---

6 The story of how Ghannam gained his status of saint is based in part on his doings as an honourable man during his lifetime. Between 150 and 200 years ago, there was a socio-political conflict about who would lead a tribe. People decided to go to Ghannam, who had previously acted as an honourable judge, to find a solution. Ghannam was not supported by a big, strong tribe, while the other people were strong and tough fighters. While Ghannam heard their pretences, one person showed no respect for him as a judge. When Ghannam realized this, he held up a piece of goat dung. He asked the audience, 'What is this?' They answered: '*Ba`ar*' (goat dung). He crushed it between his fingers, and said: 'What is this?' They said: raisins (meaning pieces). Ghannam added: 'I will crush you like this if you keep behaving in the same wrong way.' One of the audience members then mocked Ghannam saying: 'Do you think you are saint?' to which Ghannam replied: 'Oh … Please God if this secret is given to me from God … I ask you God to send the answer to what you hear and what you see.' After Ghannam had made his judgment, the man rode his horse back but accidentally touched the gunlock or trigger of his gun. The bullet killed him, thus proving that Ghannam had close ties to God.

passed the graveyard and, when milking the goats, only blood came out. Only after returning to the site did the goats produce milk, thus showing that extraordinary events are not only benevolent actions; unlike the potentiality of the good deeds, misfortune is more regularly unfolded when disrespect or forgetfulness occurs. Rather than a mocking of God, this is seen as a mocking of the potency of the saint (cf. Geary 1994: 95–115).

What matters is not the corporeality of the dead body, or even his/her actions during lifetime, although this cannot be excluded as a mechanism in the further negotiation of evidence of sainthood. What is crucial is the materiality of the grave, ground, and knots to communicate between the earthly and divine worlds. Although the practices have largely stopped, as illustrated below, some people still offer the first clarified butter (*samn*) from the goats at the saint graves, and when a new baby is born it may also undergo a form of initiation with a visit to the graves.

The saint graves offer what Giorgio Agamben has called 'the hardest and bitterest experience possible: the experience of potentiality' (1999: 178). The saint grave is a *potential* in the sense that it has, in a generic sense, transformed the person from an ordinary deceased person into a saint, through extraordinary events, but with an existing potential to also *not* actualize wishes. Extraordinary events may take place at the grave; death, it seems, is not the end. To Agamben, potentiality is thus the very possibility of *not* actualizing. It is this potentiality of acting, not acting or acting differently, that survives in oral histories among the tribes rather than a notion about healing every pilgrim, or questions of the intention of the saint, God or whatever. As Maniura aptly states, 'The characteristics of saints are to be sought not in the inherent qualities of a category of persons, but in the behaviour of those who interact, or attempt to interact, with them' (2009: 629). Through such practices, it is also materiality that people in the area are forced to deal with; the things offer a potentiality that cannot be easily ignored.

This question of potentiality has to a certain extend been explored through the concept of affordances (Gibson 1979; Ingold 2000; Knappett 2004). The concept of *affordances* leaves it up to a (living) person to establish what an object is used for within the potential framework of the design (for instance, a chair can be used as a ladder, but not as a computer). Viewed through this perspective, the 'functionality' of a thing is not the issue; rather, the question is the potential ways in which its form may be applied to a task. My point with potentiality here is, however, that the question of what an object potentially *does* is placed at the object end, at the grave, with all its attributes and connections to divine influence. It is thus the potential of a dead saint and the grave to act (or not act) in seemingly extraordinary ways, rather than living people who define what a thing can actualize for certain.

The excess of agency encountered at the saint graves becomes a powerful uncertainty that is encapsulated in Agamben's notion of 'the experience of potentiality'. It is this potentiality of agency, this excess of agency, that is unsettling to both the religious scholars and people in the area, whether they visit the graves or not – things happen, have happened and *can happen*.

New Regimes of Knowledge

Yet, excess provokes excessive response, either of worship or of persecution (Meltzer and Elsner 2009: 377). Thus it is also an 'excess' that should be handled with care (cf. Kuzmanovic in press; Bille in press). Since the 1990s the pilgrimages have largely disappeared, in light of the spread of an Islamic revival. According to the Islamic scholars in the area the pilgrimages, to Jebel Haroun in particular, were clear proof of the religious ignorance of people in the area. By relying on the saint for blessing, people were associating others with God. However, the few people who still visit argue that the many stories of success testify to the potency of the graves. Others neither believe nor disbelieve in the power of the saint and the graves, but rather abstain from judgment, suspending the whole question to God's will (Marcus 1985: 456). Even in the cases where people no longer visit the graves, they are highly critical of the new scholarly regime, as life has deteriorated since the pilgrimages stopped; there is an uncanny feeling of the saint's retaliation. This demonstrates how being offered blessings by the dead is also a dangerous debt that may become an obligation. One informant expressed this by saying:

> They [the saint worshippers] visited and came back with a pure purpose. Religion and scholarship has increased today. They [the scholars] say that visiting the tombs is forbidden and people started visiting them less and less. But when people started visiting less, things changed. How did things change? It rained less, life became harder, and the animals didn't find anything to eat. People had more faith then, than today. Nowadays, faith is weak … They say that visiting the tombs is forbidden, visiting the sanctuary of Prophet Aaron is forbidden. Is it forbidden to read the Fatiha [opening verse] for the dead? No! Some experts say that it is forbidden. It's forbidden to visit the tombs of your parents to read the Fatiha! This is wrong! My father is dear to me dead or alive and it is my duty to visit his tomb. The most important thing is the consciousness between myself and God. I don't defy God, but for my consciousness it is important for me to visit my father's tomb. And God knows that I don't disbelieve in him.

The Islamic scholars I interviewed were not against reading the opening verse at the grave, as such, only the intermediation. The informant above does, however, describe a general sentiment among the elderly, of how things have deteriorated since the pilgrimages stopped. For the pilgrims, the saint is 'a living presence and active agent in the world, existing between God and his believers and as a go-between' (Gilsenan 2000: 605) – a presence offered by the excess ejecting from the grave.

But saints are also contested figures, as they interfere with structural powers. Michael Gilsenan notes that, 'To the state and to those attempting to regulate the spaces of power and citizenship, that person is above all a source of disorder and the illusory forces of magic' (2000: 605). One such powerful place of 'illusory forces' is located between Taibeh and Wadi Mousa. The graveyard is variously known as Ayn

Figure 7.2 Saint's grave in Ayn Amuun, near Taibeh (photograph by author)

Amuun, after its location next to a spring, or al-Fuqara, after four graves belonging to a category of people known as a faqiir (plural *fuqara*) associated with spirituality, poverty and divine blessing. The graves at Ayn Amuun are positioned with a clear view of Aaron's tomb on Jebel Haroun. On one side of the new asphalt road that cuts the graveyard in two, three of the graves have small heaps of stones, where the only signs of the special status of the deceased are the marks from incense burning and the occasional gifts.[7] On the other side of the road is one of the most elaborate demarcations of saint worship in southern Jordan (Figure 7.2).

A large square wall surrounds several graves in the centre of the walled area.[8] Extensive amounts of objects are left at these graves for healing and good fortune, such as the remnants of candles and coins; there are darkened patches on the wall from the burning of incense and the ubiquitous pieces of cloth tied to a stick, which is placed in between the stones of the wall.[9] As elsewhere, stories abound of

7 These graves are mostly attributed to the e'yeanat tribe.

8 Some say that the main grave there is the grave of al-Kheisi, from who people from the cities of Dana and Tafila claim ancestry. The story I most often heard was that it was the grandfathers of the Saaidat tribe, Salman Saaidat and Salim Saaidat. According to one story, when people stopped visiting the graves they would hear stones thrown at their houses all night, without finding the culprit. It only stopped after they visited the grave again.

9 Outside the cemetery is yet another grave called *khbaybi* where incense is burned. However, this grave is not considered to belong to a saint, but rather to a cheerful person who was always spreading good humour. In hagiographies, such a distinction between saint and 'fool' is not always so clear-cut.

the healing successes, including stories of sick people who would sleep inside the structure until they were healed.

In light of the Islamic revival, these practices have mostly ceased, particularly in densely populated areas, although proof of its continued practice is still evident. This revival entails a more scriptural understanding of Islam and is spreading through mosques, televisions and schools, being particularly inspired by the Gulf countries and Saudi Arabia. Saintly intermediation is considered *bid'ah*, an innovation, but contrary to the English word it has predominantly negative connotations. In this line of religious thought, veneration of the saints is equal to polytheism, *shirk*, and should be abandoned. Praying for the dead in the afterlife is permissible, but not for the sake of healing, which is to associate others with God. It is important to note here that what my informants judge is not the status of the deceased as a saint but rather the practices of the few who continue to perform them today and the many in the past. Thus, the question is not so much whether the saint grave offers blessings rather it is to assess the engagement with this excess (although the clergy of course may question the saintliness of the person whose grave it is).

Knowing Ignorance

The disdain with which the venerating people are met is framed through the notion of *jahiliiyyah* – belonging to the age of ignorance. The term *jahiliyyah* generally has two connotations: (1) an historical epoch when pre-Islamic Arabs were primitive barbarians, considered to be ignorant, and (2) any conduct which goes against Islamic culture, morality or Islamic ways of thinking and behaving (Goldziher 1967: 201–9; Shepard 2001). Up until the twentieth century, *jahiliyyah* mostly referred to the former, a time of ignorance, of superstition and polytheism. This understanding stresses the religious state of unawareness of the guidance from God and his Oneness. The second understanding emerged in popularized form during the twentieth century with the rise of a more politically oriented Islam. This political turn may be central to the Islamic revival and the emphasis on a scriptural understanding of Islam, but only to a limited extent inspires the political developments around Petra where the theological implication and inspirations are clearer. One of the main inspirations for this Islamic revival is the Egyptian author Sayyid Qutb who describes the understanding underlying the changing perception of religious practices as follows:

> When a person embraced Islam during the time of the Prophet – peace be on him – he would immediately cut himself off from *Jahiliyyah*. [The state of ignorance of the guidance from God.] When he stepped into the circle of Islam, he would start a new life, separating himself off completely from his past life under Ignorance of the Divine Law. He would look upon the deeds of his life of Ignorance with mistrust and fear, with a feeling that these were impure and could not be tolerated in Islam! With this feeling, he would turn toward Islam

> for new guidance; and if at any time temptations overpowered him, or the old habits attracted him, or if he became lax in carrying out the injunctions of Islam, he would become restless with a sense of guilt and would feel the need to purify himself of what had happened, and would turn to the Qur'an to mould himself according to its guidance. (Qutb 1981: 30–31)

On a general level, categorizations of time periods are integrated into social life and often emerge as ruptures in the slipstream of new religious movements. It is a categorization of time with social, even psychological, implications of guilt and morality that is also entangled into social life in Petra. Death may offer a different status to a person, such as sainthood, but it is the practices of the people at the saint grave that lead to the moral judgments by those who believe that they practice a properly informed Islam. By finding their recourse in the past at the time of the Prophet, the adherents of the Islamic revival seeks to 'rid the world of magic'; a disenchantment project it shares with the advent of modernity and rationalization but which, nonetheless, as Gilsenan (2000) notes, relies on its own enchantments.

However, the power of agency represented in oral stories seems to linger with people, especially the elders, who are not so convinced by the prescriptions arising from a solely scriptural understanding of Islam. While many of my informants in their late thirties and older explicitly ridiculed people practising saintly intermediation, stories of retribution are still quite common among them. One often-told story about the graves at Ayn Amuun is about a soldier returning from Taibeh (a city north of Petra) who visited the graves to read the opening verse of the Koran. He was accompanied by a friend who called him ignorant for believing that reading to some stones would help him in any way. The next day the friend's son died, and he became convinced of the retaliatory power of the saint, and returned to the grave to slaughter a goat.

Even though the clergy has been very successful in persuading most people through the Koran and Sunnah (the practices of Prophet Mohammed) to abandon these intermediation practices, most people also recognize that 'ignorant practices' still occur to a limited extent on the margins of the heavily populated areas. The problem is that most people who scorn saintly intermediations recognize that there are indeed deceased people in this area that can be characterized as 'saints', and that it is dishonourable to accuse people of lying about the deeds of these people and events at the graves. They distance themselves from belief in the potentiality of the dead by categorizing such practices as belonging to a particular time – a time of ignorance. But by accepting the stories of success and retaliation, they also have to deal with the human and material causation through which these stories are formed. Sainthood, it seems, is articulated and dealt with, with a high degree of precariousness (Bandak and Bille in press). Most often, the adherents to this new regime of Islamic knowledge deal with the saints through indifference, treating the graves as places of tribal identification, and otherwise abstain from judgment by leaving it up to God's will. Few others deny the potentiality of the graves and accuse the saint worshippers of apostasy.

One consequence of this stance has been an animosity towards the graves of the dead saints. Some of the signs of veneration of the graves are currently being destroyed, such as the excavation of Hamad's tomb at the Hamad cemetery, where the two other graves of Ghadi Mohamed and Sabah still stand, but without the beams with knotted cloth.[10] At other places, the knots and other marks of veneration are simply disappearing, as they no longer are performed and the wind tears them down.[11] Places like the shrine at Janeb ibn Janeb, where light was said to glow at times even though no one was there, are now reduced to rubble, and shrines of Nusha bin Nusha in Abu Makhtuub or Jebel Karoun only have a few remnants of the elaborate structures left. Other sites for intercession only exist through oral traditions, rather than in material traces, and may not be centred on graves. This includes Zegnat Eshruur in Wadi Bodha near Taibeh, which was a shrub next to a grave of a female saint, the tree called Shajarat al-Minjeh with elaborate knots of cloth tied to it, and another tree called al-Tameemah near two saint graves at al-Arja cemetery.

The issue at stake here thus essentially concerns the negotiations taking place between various regimes of knowing, or rather not-knowing, in Petra. In his work on ignorance Roy Dilley notes that ignorance 'is not a simple antithesis of knowledge … It is a state which people attribute to others and is laden with moral judgement' (2010: 188). In this sense, the designation of a particular category of time in relation to how people deal with death becomes a moral mark of knowledge and piety – or lack thereof (High et al. 2012). Dealing with the deeds of a dead saint becomes subject to a social and pious positioning through claims of knowledge and ignorance.

While occurring in a radically different context, this is also what Joel Robbins (2004) refers to among the Urapmin community of Papua New Guinea, where old traditions have been replaced by Christianity. The point is that in times of rapid transformation, new moral collectivities and values also emerge. This also means that 'because people must orient to these values in their decision-making process, even if only to flaunt them, it is also a field in which naked self-interest cannot simply run the table' (Robbins 2004: 315). However, actions open to moral judgment do not necessarily include those people who do not know any better.

10 This is also the case at the Abu Khiribe cemetery on the way to Jebel Haroun from Wadi Mousa, where the sticks at the grave of faqiir Salame were intentionally crushed beneath stones. Similarly, the beams of one of the three saints' graves at the graveyard at al-Muraibit have been removed; the beam is still standing by the grave of Farraj bin Mufarraj and Mohammed Sweilim Farraj. One of the descendants told me that, on the one hand, he understood that it was un-Islamic but, on the other, he also felt that it was unjust to the deceased saint that it was removed on the basis of current religious arguments.

11 Another place of intermediation practices is at the stone of Umm Mathhur, where people crawl into a small hole beneath it, which allegedly has the capability of healing physical illnesses. Lastly, there is Mentar al-Dakheeirah, which is a large stone, also known as Mentar al-Dabusah. It stands several metres over the sloping ground on a hard compact pillar of soil that (miraculously) has not been washed away by the yearly floods.

Thus, in this case saintly intermediation performed by people who have not been sufficiently informed of proper Islamic procedures is seen as amoral, rather than immoral (2004: 315–16).

Among the younger population only few have substantial knowledge of the saint graves but they often ridicule those who practice intermediation. As Dilley eloquently notes, ignorance 'is an absence that has effects in terms of what is construed as knowledge, and of what social relations of learning are established in order to address the consequences of absence' (Dilley 2010: 188). In this case, it is a social relation of Islamic knowledge that centres on the scriptural sources of the clergy, either physically present in the area or televised. This clergy – and more importantly the way their message enters the social life of people with less Islamic education – seek to purify the Islamic practices from innovations, enchantments and the agency of a particular kind of material tradition of seeking providence against misfortune (Bille 2010).

It is thus in the formulation of sainthood that the excessive agency of the dead spills into everyday life and becomes a contested field of knowledge and practice (Meltzer and Elsner 2009: 375). This returns us to Gilsenan's point above that the agency of saints needs regulation and taming, so as not to become a source of religious, and by effect, social disorder (2000: 605). This includes taming interactions with the dead saints, and timing the practices in terms of eras of knowledge and ignorance (Harris 2007; High et al. 2012; Littlewood 2007).

Final Remarks: Taming Excess, Timing Time

The point of this exploration of extraordinary events around graves in the Petra region has been to show how places and stories of the potential agency of the dead are part of healing practices, tribal history and contestations in religious life – regardless of whether the events actually took place in any historical sense. The physical life of the saint has ended, undoubtedly, but his or her deeds are sustained, even enhanced, through the agency of the material remnants at the grave, which the devotee may – or may not – be blessed by: people never know the intentions of the saint, but they know of the potentiality of the grave to cause events to happen.

I have in this chapter pointed to three concerns in understanding the relationship between death, matter and time. Firstly, adding to the debate about 'the agency of things', much of the discussion has centred on whether 'intentionality' is integral to 'agency' or not. What I have tried to do in the above is also to think of agency in relation to death through the notion of potentiality, the potential for acting and not acting, as central to the narratives of the social life of the dead saints. The saint graves offer an excess of agency. It is not a human agency related to questions of intentionality. But neither is it solely a non-human agency based on causal, mechanical relations. Rather it is an excessive agency that spills over into life among the living by offering a potentiality. While some have rightly suggested that the question '*who* did it?' should be replaced by '*what* is happening?' (e.g.

Law and Mol 2008: 74) I suggest here with a focus on excess and potentiality that the questions are rather, 'could it happen again?' and 'how to deal with it?'

Secondly, as Dilley argues, while 'knowledge provides a sense of certainty about things, and has a reassuring effect regarding our place in the world, ignorance, by contrast, can suggest uncertainty and a discomfort about the world' (2010: 188). This uncertainty, I argue, is countered by the experience of potentiality that an enchanted world of saint graves offers by bridging between the dead and the living, between the invisible and the seen (Brown 1982: 6; Goldziher 1971: 259). This is a question of the sensuous experience of matter at the grave, and evaluating causation and potentiality in the stories by those (few) people who still practice or know of saintly intermediation, not of logic based on scriptural readings. Leaving these stories of excess behind is in a practical sense also a distancing from specific tribal history, from belonging, and from a sense of cultural heritage, and thus not unproblematic for those who have stopped going to the graves or reject their power.

Lastly, I have pointed to the way the categorization and, indeed, the very premise of time passing is configured on an individual and collective level by linking notions of time to the propriety, even piety, of dealing with the dead. Categorizing time by demarcating times of ignorance or knowledge, and thus passing moral judgment on saint worshippers, is a means of taming the potentiality of agency that emerges between the living and the dead at the saint graves.

References

Agamben, G. 1999. *Potentialities: Collected Essays in Philosophy*. Stanford: Stanford University Press.

Al-Salameen, Z. and Falahat, H. 2009. Religious Practices and Beliefs in Wadi Mousa Between Late 19th and Early 20th Centuries. *Jordan Journal of History and Archaeology*, 3(3), 175–204.

Bandak, A. and Bille, M. (eds) In press. *Politics of Worship in Contemporary Middle East: Sainthood in Fragile States*. Leiden: Brill.

Bille, M. 2010. Seeking Providence Through Things: The Words of God versus Black Cumin. In M. Bille, F. Hastrup and T.F. Sørensen (eds), *An Anthropology of Absence: Materializations of Transcendence and Loss*. New York: Springer Press, 167–84.

Bille, M. 2012. Assembling Heritage: Investigating the UNESCO Proclamation of Bedouin Intangible Heritage in Jordan. *International Journal of Heritage Studies*, 18(2), 107–23.

Bille, M. In press. The Samer, the Saint, and the Shaman: Ordering Bedouin Heritage in Jordan. In A. Bandak and M. Bille (eds), *Politics of Worship in Contemporary Middle East: Sainthood in Fragile States*. Leiden: Brill.

Brown, P. 1981. *The Cult of the Saints: Its Rise and Function in Latin Christianity*. Chicago: University of Chicago Press.

Brown, P. 1982. *Society and the Holy in Late Antiquity*. Berkeley: University of California Press.

Canaan, T. 1929. Studies in the Topography and Folklore of Petra. *Journal of Palestine Oriental Studies*, 9, 136–218.

Canaan, T. 1930. Additions to 'Studies in the Topography and Folklore of Petra'. *Journal of Palestine Oriental Studies*, 10, 178–80.

Cunningham, R.B. and Sarayrah, Y.K. 1993. *Wasta: The Hidden Force in the Middle Eastern Society*. Westport: Praeger.

Denny, F.M. 1988. 'God's Friends': The Sanctity of Persons in Islam. In R. Kieckhefer and G.D. Bond (eds), *Sainthood: Its Manifestations in World Religions*. Berkeley: University of California Press, 69–97.

Dilley, R. 2010. Reflections on Knowledge Practices and the Problem of Ignorance. *Journal of the Royal Anthropological Institute*, 16(1), 176–92.

Frazer, J.G.S. 1993 [1922]. *The Golden Bough: A Study in Magic and Religion*. Ware: Wordsworth.

Geary, P.J. 1994. *Living with the Dead in the Middle Ages*. Ithaca: Cornell University Press.

Geertz, C. 1968. *Islam Observed*. New Haven: Yale University Press.

Gell, A. 1992. The Technology of Enchantment and the Enchantment of Technology. In J. Coote and A. Shelton (eds), *Anthropology, Art and Aesthetics*. Oxford: Clarendon Press, 40–63.

Gell, A. 1998. *Art and Agency: An Anthropological Theory*. Oxford: Oxford University Press.

Gellner, E. 1963. Saints of the Atlas. In J. Pitt-Rivers (ed.), *Mediterranean Countrymen: Essays in the Social Anthropology of the Mediterranean*. Paris: Mouton & Co., 145–57.

Gellner, E. 1981. *Muslim Society*. Cambridge: Cambridge University Press.

Gibson, J.J. 1979. *The Ecological Approach to Visual Perception*. Boston: Houghton Mifflin.

Gilsenan, M. 2000. Signs of Truth: Enchantment, Modernity and the Dreams of Peasant Women. *Journal of the Royal Anthropological Institute*, 6, 597–615.

Goldziher, I. 1967. *Muslim Studies. Vol 1*. Albany: State University of New York Press.

Goldziher, I. 1971. *Muslim Studies. Vol 2*. Chicago: Aldine,

Hallam, E. and Hockey, J. 2001. *Death, Memory and Material Culture*. New York: Berg.

Hammoudi, A. 1997. *Master and Disciple: The Cultural Foundations of Moroccan Authoritarianism*. Chicago: University of Chicago Press.

Harper, S. 2010. The Social Agency of Dead Bodies. *Mortality*, 15(4), 308–22.

Harris, M. (ed.) 2007. *Ways of Knowing: Anthropological Approaches to Crafting Experience and Knowledge*. Oxford: Berghahn.

Hertz, R. 1960 [1907]. *Death and the Right Hand*. Aberdeen: Cohen and West.

High, C., Kelly, A.H. and Mair, J. (eds) 2012. *The Anthropology of Ignorance: An Ethnographic Approach*. New York: Palgrave Macmillan.

Hirschkind, C. 2006. *The Ethical Soundscape: Cassette Sermons and Islamic Counterpublics*. New York: Columbia University Press.

Hockey, J.L., Komaromy, C. and Woodthorpe, K. 2010. *The Matter of Death: Space, Place and Materiality*. Basingstoke: Palgrave Macmillan.

Ihde, D. 1990. *Technology and the Lifeworld*. Bloomington: Indiana University Press.

Ingold, T. 2000. *The Perception of the Environment: Essays in Livelihood, Dwelling and Skill*. London: Routledge.

Johannsen, N. 2012. Archaeology and the Inanimate Agency Proposition: A Critique and a Suggestion. In N. Johannsen, M.D. Jessen and H.J. Jensen (eds), *Excavating the Mind: Cross-Sections through Culture, Cognition and Materiality*. Aarhus: Aarhus University Press, 305–47.

Knappett, C. 2004. The Affordances of Things: A Post-Gibsonian Perspective on the Relationality of Mind and Matter. In E. Demarrais, C. Gosden and C. Renfrew (eds), *Rethinking Materiality: The Engagement of Mind with the Material World*. Cambridge: McDonald Institute for Archaeological Research, 43–51.

Knappett, C. and Malafouris, L. (eds) 2008. *Material Agency: Towards a Non-Anthropocentric Approach*. New York: Springer.

Kuzmanovic, D. In press. Imbued with Agency: Contesting Notions of the Extraordinariness of Türkan Saylan. In A. Bandak and M. Bille (eds), *Politics of Worship in Contemporary Middle East: Sainthood in Fragile States*. Leiden: Brill.

Laet, M.D. and Mol, A. 2000. The Zimbabwe Bush Pump: Mechanics of a Fluid Technology. *Social Studies of Science*, 30(2), 225–63.

Latour, B. 1992. Where are the Missing Masses? The Sociology of a Few Mundane Artifacts. In W.E. Bijker and J. Law (eds), *Shaping Technology/Building Society*. Cambridge, MA: MIT Press, 225–58.

Latour, B. 2005. *Reassembling the Social: An Introduction to Actor-Network-Theory*. Oxford: Oxford University Press.

Law, J. and Mol, A. 2008. The Actor-Enacted: Cumbrian Sheep in 2001. In C. Knappett and L. Malafouris (eds), *Material Agency: Towards a Non-Anthropocentric Approach*. New York: Springer, 57–77.

Littlewood, R. 2007. *On Knowing and Not Knowing in the Anthropology of Medicine*. Walnut Creek: Left Coast.

Mahmood, S. 2005. *Politics of Piety: The Islamic Revival and the Feminist Subject*. New Jersey: Princeton Press.

Makris, G.P. 2006. *Islam in the Middle East: A Living Tradition*. Malden: Blackwell.

Malafouris, L. 2008. At the Potter's Wheel: An Argument for Material Agency. In C. Knappett and L. Malafouris (eds), *Material Agency: Towards a Non-Anthropocentric Approach*. New York: Springer, 19–36.

Maniura, R. 2009. The Absent Saint. *Critical Inquiry*, 35, 629–54.

Marcus, M.A. 1985. 'The Saint Has Been Stolen': Sanctity and Social Change in a Tribe of Eastern Morocco. *American Ethnologist*, 12(3), 455–67.
Meltzer, F. and Elsner, J. 2009. Introduction: Holy by Special Application. *Critical Inquiry*, 35, 375–9.
Meri, J.W. 2002. *The Cult of Saints among Muslims and Jews in Medieval Syria*. Oxford: Oxford University Press.
Meyer, B. and Moors, A. (eds) 2006. *Religion, Media, and the Public Sphere*. Bloomington: Indiana University Press.
Mol, A. 2002. *The Body Multiple: Ontology in Medical Practice*. Durham, NC: Duke University Press.
Oxford English Dictionary. 2010. Edited by Angus Stevenson, Oxford University Press. Oxford Reference Online. Oxford University Press. Copenhagen University Library. 10 August 2012, http://www.oxfordreference.com/views/ENTRY.html?subview=Main&entry=t140.e0013260.
Qutb, S. 1981. *Milestones*. Delhi: Markazi Maktaba Islami.
Robb, J. 2004. The Extended Artefact and the Monumental Economy: A Methodology for Material Agency. In E. Demarrais, C. Gosden and C. Renfrew (eds), *Retinking Materiality: The Engagement of Mind with the Material World*. Cambridge: MacDonald Institute for Archaeological Research, 131–9.
Robbins, J. 2004. *Becoming Sinners: Christianity and Moral Torment in a Papua New Guinea Society*. Berkeley: University of California Press.
Roberts, A.F. and Roberts, M.N. 2003. *A Saint in the City: Sufi Arts of Urban Senegal*. Los Angeles: UCLA Fowler Museum of Cultural History.
Shenoda, A. 2008. *(Re)Producing Holiness: Coptic Christians, Mass Media, and the Miraculous*. Presented at American Anthropological Association, San Francisco, CA (22 November).
Shepard, W.E. 2001. Age of Ignorance. In J.D. McAuliffe (ed.), *Encyclopaedia of the Quran*. Leiden: Brill, 37–40.
Sørensen, T.F. 2009. The Presence of the Dead: Cemeteries, Cremation and the Staging of Non-Place. *Journal of Social Archaeology*, 9(1), 110–35.
Stauth, G. (ed.) 2004. *On Archaeology of Sainthood and Local Spirituality in Islam: Past and Present Crossroads of Events and Ideas*. Bielefeld: transcript Verlag.
Stewart, D.J. 2001. Blessing. In J.D. McAuliffe (ed.), *Encyclopaedia of the Quran*. Leiden: Brill, 236–7.
Verbeek, P.-P. 2008. Morality in Design: Design Ethics and the Morality of Technological Artifacts. In O.E. Vermass, P. Kroes, A. Light and S. Moore (eds), *Philosophy and Design: From Engineering to Architecture*. Berlin: Springer, 91–103.

Chapter 8
Pilgrimage Landscape as Material Anchor of Time

Jesper Østergaard

Introduction

According to local Buddhist tradition, transempirical mythological events are embedded in the landscape of the pilgrimage site of Maratika in Nepal, and human actors connect to these events through bodily interaction with the geographical location.[1] The landscape around the Buddhist and Hindu pilgrimage site of Maratika is considered by visiting Buddhist pilgrims to be the location where the religious figure Padmasambhava in mythic time subdued demons, removed obstacles and achieved immortality. The landscape materially express these believed events: a rock and stalagmites are considered to be the body, heart and intestines of a demoness, depressions in a cave the footprints of Padmasambhava and the landscape in total is considered today to be blessed and thus to bestow longevity. Pilgrims approach the empowered landscape in their ritual activity, they circumambulate hills, they squeeze their bodies through narrow passages and they eat the soil, thus using material substances and the physicality of landscape. Through these ritual practices they establish a relation between the immortality of Padmasambhava and their own life and death. Two different temporalities meet in the landscape: the mythological time of religious heroes and the human life time. This chapter explores the relation between the immaterial, mythological time in which Padmasambhava was believed to achieve immortality and the material, perceptual space through which human beings, who eventually will die, can achieve empowerments that bestow them with long life. Karma introduces a third kind of time, namely the circular karmic time of birth, death and rebirth to which all living beings are subject to according to Buddhist belief. By interacting with the material landscape of Maratika, pilgrims negotiate religious ideas about immortality, long life, and their own death; further through various ritual technologies they enroll themselves in the karmic relationality inherent in reincarnation. In this pilgrimage,

1 Parts of this chapter are rewritten sections from my PhD dissertation *The Topographic Mind*, where the written materials and the collected field materials are analyzed from a cognitive perspective (Østergaard 2011). This chapter is based on readings of the limited literary materials on Maratika and field observation during the Tibetan New Year celebration in 2009.

time on the ultimate scale (religious heroes' transgression of time through their enlightenment), the large scale (the reincarnation system) and on a micro scale (each pilgrim's individual life) is connected in the ritual reality at Maratika.

These different conceptions of time are thus conflated in the sacred landscape of Maratika through the rituals. It will be illustrated how the pilgrimage landscape functions as a material bridge between mythological time, karmic time and human life time by analyzing how material landscape anchors the immaterial times in material spaces. Empirical topics such as the narrative representations of the landscape of Maratika in a Buddhist religious guidebook and pilgrims' ritual integration of landscape will also be examined.

Time in Maratika

In religious traditions, immaterial concepts of time are often materialized in/as symbols, paintings, statues and temples. Such material artifacts bring physical form to immaterial cultural realities: they enable perceptual interaction with cultural values, standardized in cultural models (Day 2004). One famous representation of time in Buddhism is the "Wheel of Life" (*srid pa'i 'khor lo*),[2] illustrating how karmic time covers six realms of existence (god, semi-god, human, animal, ghost and hell), one of which includes the human life time. Enlightened beings, such as Padmasambhava, since they have realized the illusoric nature of Samsara, transgress this Samsaric wheel into a nirvanic or mythological "time," which more correctly is considered to be beyond any kind of time. According to Tibetan Buddhism such enlightened, religious heroes re-enter Samsara in order to help other beings in their human life time and in their karmic time and ultimately to become enlightened. Mythological time is therefore also the end of time when all beings are enlightened and liberated from karmic existence, thus mythological time frames other times. The Wheel of Life illustrates, among other ideas, how human life time that is embedded in karmic time is transgressed by ultimate mythological, enlightened time, thus, when human life time is interrupted by death, karmic time takes over until mythological time is achieved through enlightenment by the help of religious heroes. These three levels of time correspond to three different orientations in Tibetan Buddhism, the Bodhi (enlightenment) orientation, where the goal is to escape repeated rebirth in Samsara, the karmic orientation, where the goal is to ensure a favorable (or avoid an unfavorable) rebirth and a pragmatic orientation, concerning enhancement of human life (Samuel 1993: 159).

The three different times are practiced upon and negotiated in Maratika, where mythological, karmic, and human life time intertwines. Humans are related to those three times, just as the times are embedded in Maratika as a sacred place. Changes in space are also, in this framework, changes in different times and interactions

2 Central words and names are also given in Tibetan, using the Wylie-system of transliteration.

between times. The materiality of landscape enables pilgrims to bodily and perceptually negotiate their human life time and its inevitable end. A pilgrimage to Maratika thus negotiates pilgrims' own death by utilizing material substances of landscape as time capsules that enables pilgrims to leap between mythological, karmic and human life time. Pilgrim because being human cannot be at two places in one time, but on a visit to Maratika be in two (or more) times at one place.

Material Anchors of Time

By giving materiality to cultural, conceptual matters, illustrations such as the Wheel of Life act as "material anchors" enabling individuals and groups to share reference when dealing with religious ideas (Brandt 2005; Hutchins 2005). Religious ideas are cultural models that "are not only ideas that reside inside minds, they are often also embodied in material artifacts" (Hutchins 2005: 1558–9). Material anchors therefore stabilize religious ideas and cognition through these ideas, and thereby cognition can be expanded and become more complex: "The complexity of the manipulations of structure can be increased if the stability of the representations can be increased" (Hutchins 2005: 1557). As a consequence, religious representations of socially constructed times, although used in individual timework, are often external, materialized time tools in religious artifacts, and situated in a shared, social world. External representation are prevalent in religion since "the strategic use of structured environment (e.g., 'sacred space') and external representations (e.g., statues or likeness of the gods) could put individual agents in a position to better utilize their basic perceptual, motor, and social cognitive skills in these domains" (Day 2004: 117).

Material anchors are prominent tools as external representations of time. The face and hands on a watch are extensions in physical space, and exactly because they give material spatiality to chronological time, the watch gives the user the possibility to "watch" time. A watch is spatialized time. By using the watch as a material anchor, the difficult task of knowing the time is transferred into a perceptual interaction between the cognizer and the watch, thus releasing cognitive resources for other cognitive tasks.[3] Instead of using cognitive energy to figure out what time it is, we can think about how to use this information for relevant purposes. Sacred spaces can be external representations of mythological events, since for the pilgrims the spatial landscape offers perceptual interaction

3 An even more radical suggestion is that the material anchors not only give more precise answers and easy solutions to difficult cognitive problems, they may themselves have generated the problems. To take the example of watches again, only with some precise measurements of time, the problem of timekeeping could arise. Indeed the usefulness of external, material anchors and other cognitive props may rest exactly in their capability to enable otherwise impossible or difficult cognition. For an overview of positions within situated and extended cognition, see Robbins and Aydede (2009).

Figure 8.1 Maratika

Maratika is situated in eastern Nepal around 140 km east of Kathmandu and 92 km southwest of Mt. Everest. In the foothills of the Himalayas, Maratika is an area of three hills with a number of caves in an altitude a little over 1,300 meters. The small surrounding village is predominantly inhabited by Hindus. According to Hindu tradition, a stalagmite in the central cave is a manifestation of a Shiva lingam (Buffetrille 1994: 3). Maratika's relation to Tibetan Buddhist history is through its connection to the life story of the Indian guru Padmasambhava.

with an otherwise inaccessible mythological time. The ritual interaction pilgrims perform with landscape are thus not symbolic actions with a symbolic landscape, but bodily actions with materialized mythology—just as two hands pointing at "12" on the face of a watch is processed by our cognitive system that the time *is* 12 o'clock, *not* symbolizing time.

Returning to landscape, its materialization and temporalization are two sides of the same coin, or in the words of Barbara Bender: "*Landscape is time materialized.* Or, better, *Landscape is time materializing*: landscapes, like time, never stand still" (Bender 2002: 103, emphasis in original). The landscape of Maratika is not referring to mythological events, it is a spatial continuation, a continuation established through religious narratives about Padmasambhava's ritual acts in Maratika. Maratika is not a memorial site to which pilgrims can turn to and remember Padmasambhava, instead the landscape of Maratika is the reminiscent of him, his trace, that like relics possess some of his ascribed qualities, like the power to manipulate karmic time and extend human life time.

The Landscape of Maratika

Maratika is known for its caves. One of these caves, the Long Life Cave, is often synonymous with Maratika and considered the most sacred cave and therefore the most frequently visited cave in the Maratika area, by both Hindu and Buddhist pilgrims. Long Life Cave is the dripstone cave where Padmasambhava allegedly meditated and achieved immortality and is the center of ritual activity. A few minutes' walk will take the pilgrim around the cave, but most pilgrims stay for longer to make offerings, touch geological features or pass through narrow passages in the rock, as will be discussed below. From early morning until late at night, the Long Life Cave is visited by pilgrims. The number of geological features related to myth is great; mentioned here are only those considered most important. The Long Life Vase (*tshe bum*) is the most revered geological object in Maratika by Buddhists, who will touch it with hand or forehead, emphasizing the connection between the materiality of landscape and the body of pilgrims. In the narratives, it is stated how the Buddha of Long Life, Amitayus (*tshe dpag med*), placed the Long Life Vase in the cave as a testimony to Padmasambhava and his ritual helper Mandarava and their ritual activity in Maratika. Behind the Long Life Vase, some furrows on the wall of the cave are considered to be Dakinis (female deities) writings. Different stalagmite formations resemble other mythological figures, like the Four Guardian Kings (*rgyal chen bzhi*) that are protectors of the four cardinal directions of the world and often found depicted at the entrance to temples and a "wish-fulfilling gem" (*'dod g.yang gi ga'u*) that, according to a belief shared by pilgrims, will grant offspring and possible wealth to the devoted pilgrim when they make offerings, e.g. grains, to it. Pilgrims collect pebbles or grains from below the stone. This is likewise a simple, divinatory practice: a whole grain of rice indicates a boy, whereas a broken one indicates a girl as the sex of the offspring. Below the Long Life Vase there is a large impression in the wall of the rock considered to be a body print of Amitayus. On the same wall of the cave, a newly discovered material manifestation of myth is the "Guru Rinpoche [i.e. Padmasambhava] Body Trace" on the wall of the cave just below the Long Life Vase. Formations in the dripstones are considered by the Buddhist lama, Karma Wangchuk, to be a manifestation of Padmasambhava and thus he could draw an image of Padmasambhava along the lines of dripstones on a photo of this part of the cave (Østergaard 2010, 2011).

Placing Mythological Time in Maratika

The description of the special features of the landscape is often found in oral and/or written tradition. As in many other sacred places (*gnas*), religious agents wrote a religious "guidebook" (*gnas yig, gnas bshad*) to Maratika. The 15-page "Introduction to Maratika" was written by the founder of the monastery, Ngawang Chöphel (Ngawang Chöphel 2000b) in the mid 1990s and is now included in *The*

Figure 8.2 Bhutanese pilgrims below the Long Life Vase at the top of the cave. Behind the vase, seams in rocks are believed to be Dakini writings

Guide to Maratika (Karma Wangchuk 2000), a small booklet published by the Buddhist monastery at Maratika. This publication is a compilation of different texts in Tibetan, with English translations, concerning Maratika as a Buddhist pilgrimage site and the founder of the monastery, Ngawang Chöphel. Maratika is the place where Padmasambhava,[4] according to "Introduction to Maratika" and believers, achieved immortality (*'chi med*). Maratika, a place in the geographical world, gains religious importance through these assumed mythical events. Where myth and landscape meets, a sacred place (*gnas*) is established. A sacred place thus conflates the materiality of the landscape with religious significance, and indeed the Tibetologist Toni Huber states that: "*gnas* … always carries the double meaning of the actual physical place, and of the residence or existence of deities, entities or beings believed to be powerful or significant in some way by the pilgrims who go there for an encounter … There is much ritual evidence to suggest that the physical 'stuff' of the place and the vital principle or being that resides there are always so closely associated that they are considered and treated as identical" (Huber 1999b: 83). It is stated by Katia Buffetrille, "The objective of a pilgrimage guide is to lead the ordinary pilgrim from the simple perception of a physical landscape to the conception of the place as a sacred landscape" (Buffetrille 1998: 21).

The Sacred History of Maratika

Pilgrimage guides have an impact on pilgrims' minds and perception of landscape, through a mythologization of landscape (McCutcheon 2000) that includes a timescaping whereby mythological time is spatialized in landscape. The sacred history of Maratika, unfolded in the guidebook, has a tripartite structure and "descends," (1) from mytho-cosmological sphere of deities and Buddhas; (2) the acts of superhuman agents, especially Padmasambhava, on earth; and (3) religio-historical concerns about the pilgrimage site. In Figure 8.3, I have illustrated how this spatialization of mythological time includes a descent from cosmology, through mythology to history and is complimented by ascent of the landscape from "nature" to sacred place. The spatialization of myth entails the mythologization or temporalization of landscape.

---

4 He is often referred to as Guru Rinpoche, "precious teacher" in Tibetan. "Rinpoche" is an honorific term commonly given to reincarnated lamas in Tibetan Buddhism. The historicity of Padmasambhava as a person is questioned (Huber 2008: 238). Padmasambhava and Avalokiteshvara (Tibetan: *spyan ras gzigs*) are the most important superhuman agents in Tibetan Buddhism. Although aspects of opposition appear in the traditions, with Avalokiteshvara considered the patron deity of Tibet (Kapstein 2000: 147), and Padmasambhava representing a foreign, powerful religious tradition, they can also be interpreted as complementing each other (Kapstein 2000: 155). They are both highly revered mythological figures throughout the Tibetan Buddhist area.

Placing Immortality in Maratika

In "Introduction to Maratika" it is first told how Amitabha (*od dpag med*), the Buddha of Limitless Light, taught 18 tantras of immortality at Avalokiteshvara's request.[5] The teaching was recorded and transcribed on gold paper, with ink of lapis lazuli, into the language and writings of the Dakini, put into a small box of five jewels and then hidden in the cave of Maratika by the Dakini Sangwa Yeshe (Ngawang Chöphel 2000b: 25–6). Concealment of teachings in the landscape for future discovery is known from the Tibetan *terma*-tradition (*gter lugs*).[6] The *terma*-tradition is especially connected to Padmasambhava, who is said to have hidden many teachings in the soil, thus the physical landscape is a potential reservoir of religious teachings throughout the area of Tibetan Buddhism—all parts of landscape are potentially sacred places. The "Introduction to Maratika" makes use of the same topoi that certain places in the landscape are mythscapes prepared for the Buddhist teaching.

Padmasambhava and Mandarava in Maratika

The second episode in the sacred history of Maratika concerns what is considered to be the most important mythological event in the history of Maratika, namely Padmasambhava achieving immortality. In order to achieve immortality Padmasambhava came to Maratika, where he first tamed the demoness residing at Maratika. According to myth, he suppressed the demoness that is now believed to be manifest as a stone in the entrance to Eight Heruka Cave, where stones and stalagmites are said to be her body, heart and intestines (Ngawang Chöphel 2000b: 37–8). Furthermore, Padmasambhava needed a ritual helper, a consort, for the practice and a few pages in the "Introduction to Maratika" describe how Mandarava leaves her home, and has horrifying experiences, walking through deserted ghost towns and cemeteries, before she meets Padmasambhava, who takes her to Maratika (Ngawang Chöphel 2000b: 27–9). During their meditative rituals together, it is said that different deities appeared before them. Essentially, the text claims that Padmasambhava meditated together with his consort Mandarava

5 The two, Padmasambhava and Avalokiteshvara (Chenrezig; *spyan ras gzigs*) are the most important superhuman agents in Tibetan Buddhism. Although aspects of opposition appear in the traditions, with Avalokiteshvara considered the patron deity of Tibet (Kapstein 2000: 147), they can also be interpreted as complementing each other (Kapstein 2000: 155).

6 The tradition goes back to tenth-century Tibet (Gyatso 1996: 148). In the *terma*-tradition of hidden "treasures" (*gter ma*), it is assumed that texts and material objects can be hidden at one time and later discovered by a treasure finder, a *tertön* (*gter ston*) (Aris 1988; Doctor 2005; Thondup 1997). There are two primary modes of discovery. One is the unearthing of treasures hidden in a geographical location (*sa gter*), another is to retrieve a treasure from one's mind (*dgongs gter*) (Gyatso 1996: 148). The Dzogchen teacher Namkhai Norbu is believed to have received a treasure through dreams at Maratika (Namkhai Norbu 1992).

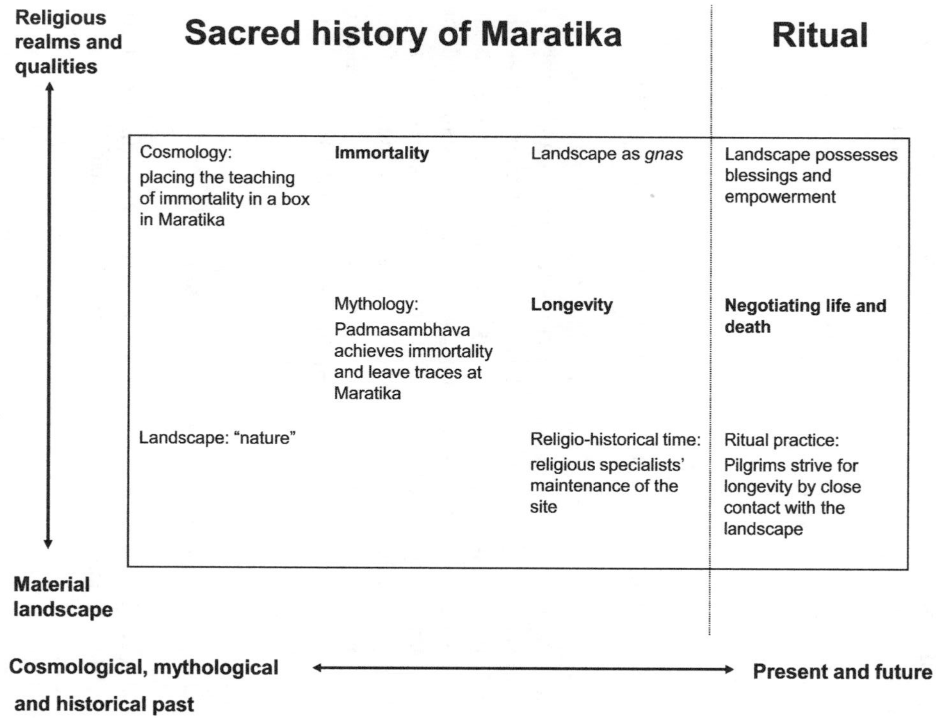

Figure 8.3 The Sacred History of Maratika

The history of Maratika as it unfolds in the guidebook descends from a purely cosmological realm (top-left), through an intersection of mythology and landscape (center) and continues in landscape (bottom-right) and approached by pilgrims (down in ritual). Immortality, long life and negotiation of death follow a similar pattern. Landscape is changed in this process, from "wild nature" (bottom-left) through a taming-process described in mythology (center) to becoming a sacred site, *gnas* (top-right) and can bestow blessing/empowerment, *byin rlabs*, on pilgrims (top in ritual) for three months, and they achieved immortality at Maratika (Ngawang Chöphel 2000b: 29). Padmasambhava's alleged ritual activities at Maratika are the most important events in the history of the place, and it is mainly this part of the story or narrative of Padmasambhava that is believed to be materialized in the cave and surroundings of Maratika as dripstone formations, rock formations and depression in rocks that pilgrims ritually interact with by touching, squeezing through and/or eating, thus placing their life in mythology. By situating mythological events in the physical landscape of Maratika, the dichotomy of myth and landscape is bridged, thus giving space to mythology and letting the mythology of Padmasambhava continuously unfold in landscape. It establishes Maratika as a mythscape where mythological time can be approached and utilized in pilgrims' taming of the inevitable course toward the end of their human life time.

Historical Developments

The third period in the sacred history of Maratika is a transition in level from a predominantly mythological time to a historical level, where the historical events are framed within a religious interpretation. It is also a shift from superhuman agents, like Padmasambhava, to religious specialists and pilgrims as the main actors at Maratika. These actors are first and foremost caretakers and visitor of the sacred place (*gnas*) of Maratika and the developments in time take place in the physical world. The recent history of Maratika is closely connected to the life of Ngawang Chöphel, a dominant religious specialist who established the Buddhist monastic community in Maratika in 1980. In his biography (*rnam thar*), included in *The Guide to Maratika*, it is stated how "he has not spared his efforts: working with enthusiasm, he has built a temple and cells [i.e. small rooms for meditation] in this sacred spot" (citation from Buffetrille 1994: 34; Ngawang Chöphel 2000a: 49–50).

Pilgrimage Landscape as Material Anchor of Mythological Events

On a mythological level, the religious significance of Maratika is still considered to be unfolding. Most important for pilgrims, however, is that landscape is empowered by the act of Padmasambhava and thus enabled to extend human life and override the chains of cause and effect of karmic time. Through the Buddhist mythologization wild nature is tamed and domesticated by Padmasambhava.[7] The

7 This process, whereby a "wild" landscape, inhabited and controlled by forces that are not in favor of giving away their powers and control of the place, are "tamed" (*'dul ba*), is referred to as "mandalization" (Grapard 1982: 207) or "Buddhicisation" (Buffetrille 1998). Then the land is "cultivated," often by placing a Buddhist conceptual map of space, in the form of a mandala, there in order to make the land fit into a widely used spatial schematization, a process in the anthropological work on Tibetan pilgrimage.

landscape at a pilgrimage site in Maratika, as in other Tibetan Buddhist sacred sites, is considered to be a geological surface imbued with a mythological depth. The pilgrim approaches the pilgrimage site not just as a set of ordinary geological features but as a set of features to which links with significant mythological topics and events are ascribed. These landscape features do not just *refer* to mythological topics and events; instead they are material anchors believed to *materialize*, *express* and *visualize* mythology. By materialized mythology, I emphasize that mythology is not only a narrative sequence of events. Because the landscape around the pilgrimage site is believed to be developing and self-produced (*rang 'byung*), these events may also be embedded and developed further in the landscape as changing geological features. Pilgrims and religious specialists referred to unusual geological features in Maratika as self-produced and naturally arising, as opposed to man-made or through human contrivance, and that, like other places, it "is a special type of place, one that is 'self-produced' … thus possessing an innate, natural power or 'empowerment' (*byin rlabs*) from which it gains its high status" (Huber 1999a: 61).

The intersection of mythological time with geographical space leads to an "artefication" of an otherwise natural landscape. By embedding mythology in the landscape, a temporal dimension is given to the landscape, i.e. landscape is temporalized. Likewise, the landscape gives material form to a time-related mythology, i.e. mythology is spatialized. Space and time(s) thus mutually reinforce each other in the landscape. Mythological time is placed in landscape through religious narratives, such as the mentioned guidebook, that creates places in the landscape as "timeless," sacred places, imbued with material substances possessing the qualities of time shifting. Because mythological time is transgressing karmic and human life time, substances that are material anchors for mythological time work on karmic and human life times.

Landscape itself, in this interpretation, becomes imbued with religious qualities, such as the ability to bestow blessings and long life, both described in the guidebook and accepted by pilgrims.

From the perspective of the religious believers, when Padmasambhava allegedly left traces behind, he also left religious qualities in the landscape. Thus, the mythology is not only or necessarily approached through the guidebook but likewise through the landscape. Therefore, I suggest, we need to look at more than just the mental representations of myth, because the religious ideas are not only *about* a landscape with believed religious qualities. In this case the landscape of Maratika, it is a shared landscape with believed religious qualities, and religious ideas gained and transmitted *through* the perceptual interaction with this mythologized and enabled landscape thus have a shared reference to an actual geographical location.

Pilgrims' Ritual Interaction with Maratika

Old carvings of mantras (magical spellings recited by pilgrims and often carved on rocks) and their attrition suggest that pilgrims have visited Maratika for decades.

The textual references to Maratika, however, are much older than the geographical location. In texts, Maratika is referred to in twelfth-century mythologies of Padmasambhava. Maratika was time before it became space. The identification of the mythological time of Padmasambhava in Maratika and the geographical location of Maratika in Nepal is thus still negotiated (Østergaard 2010, 2011). Geological features, like rock outcrops, depressions in stone and patterns in dripstones, were examined by the head of the Buddhist monastery in Maratika, Karma Wangchuk, and other religious specialists, to establish their relation to mythology. If established, and authorized by religious authorities, like Karma Wangchuk's own teacher, Trulshik Ngawang Chökyi Lodrö, then the geological features can be included in the group of religiously significant geological features as a social technology of time. Once the physical landscape is identified as a mythological landscape and authorized by religious authorities, the landscape is believed to be empowered through the achievements of Padmasambhava. The empowerment is materialized in the landscape of Maratika, and some pilgrims spend considerable time collecting the water dripping from stalactites or collecting dust from a small cave at the top of Avalokiteshvara Hill. The empowered soil and water can then be taken home and/or given to others, whenever necessary, such as in times of illness. These substances provide contact to Padmasambhava. On the packages of Long Life Pills[8] it is noted that the pills connect to Padmasambhava and have curative power. Mythological time, with its connection to a nirvanic no-time, and human life time interrupted by death, merges in these pills. An exchange of substances occurs between defilements of the pilgrim (preventing one from a long life and also giving one an unfavorable rebirth in karmic time) and the blessing qualities or empowerment of Maratika (connected to mythological time with soteriological qualities).

Pilgrims' Practices: Approaching Long Life, Negotiating Death

A variety of ritual activities can be observed among Buddhist pilgrims at Maratika. Pilgrims circumambulate Avalokiteshvara Hill, make offerings, squeeze through narrow passages in the rocks, and perform other ritual acts; some meditate, many collect pebbles, dust and water. At many places at Maratika, pilgrims touch topological features with forehead and hands. These activities often place the body of the pilgrim and the landscape of Maratika at the center of ritual interaction. In general, "[g]uiding the movement of the human body through territory has long been one of the most important roles of religion" (Corrigan 2008: 166). Guiding the body through landscape facilitates a contact between the mind of the pilgrim and the Buddhist mythology of Maratika. Ritual practices concerning the life and

8 Long Life Pills are said to liberate through tasting, as described by Gayley (Gayley 2007). In Maratika, they are made from soil, plants and water of the area and consecrated through a ritual before the *Losar* celebration and given to pilgrims on request. For such a realization of medicine ritual (*sman sgrub*), see Laurent Pordié (Pordié 2008) and for the relations between the medical tradition and the Nyingma-tradition, see Frances Garrett (Garrett 2009).

the death of pilgrims, however, not only take place in landscape, they gain efficacy through landscape. That is material substances have soteriological qualities. Ritual practices make use of the landscape of Maratika as liberating through touching or wearing (*btags grol*) and—like other Buddhist sacra—can "save beings from the karmic consequences of their own negative deeds" (Gayley 2007: 460). A human body inevitably leads to new karma and thus immortality is impossible by ordinary humans, because they are infested with karmic residue and thus not in a position to achieve immortality. At Maratika, every pilgrim "knows that one should ask for long life at Halase [i.e. Maratika]. The original event determines the pilgrim's request. Padmasambhava obtained immortality, but the pilgrim will be content to ask for a long life, aware that his spiritual development does not allow him to expect more" (Buffetrille 1994: 45). Once the human life time stops at death, karmic time takes over, but a visit to Maratika reduces pilgrims' karmic residue lowering the risk of an unfavorable rebirth as animal, as a ghost, or in hell.

Long Life Pills

Long Life Pills produced in Maratika cure, according to the description on the package, illnesses connected to "untimely death," that is, illnesses preventing one from living a full human life time and dying in old age. The Long Life Pills combine the connection to Padmasambhava and the alleged healing qualities into small round pills made from the mud, herbs and soil from Maratika and consecrated through a ritual before the New Year celebration and given to pilgrims on request. These magical pills combine medical qualities, concerning human life time, and soteriological qualities, concerning mythological time, since the empowerment (*byin rlabs*) of Padmasambhava may eventually lead to enlightenment, thus ultimately transgressing time, and curing illnesses, thus prolonging human lived time.[9] Long Life Pills are socially constructed devices using technologies of time, by conflating mythological time with human lived time.

Like Buddhist relics (Strong 2004), the pills makes qualities once located at one place or in one person distributable and moveable. Because myth is anchored in Maratika as material representation, it can be moved around as material. The believed benefits from Padmasambhava's ritual action are rolled into small pills that bring past mythological events and time into the present of pilgrims, who can take these pills home and use them with the future results—postponing death. By eating these pills, or alternatively dust, herbs or drinking water from Maratika, pilgrims continue the process: myth is placed in landscape and through landscape myth is placed within humans.

9 Pilgrims not only visit Maratika to prolong life, but also to create life through procreation, as pilgrims come to ask for a child, as could be observed and is mentioned in the biography of Ngawang Chöphel (Ngawang Chöphel 2000a).

Four Secret Paths

Despite material substances that prolong life, a human body eventually will die and reincarnation is inescapably connected to karma according to Buddhism. Because death puts an end to human life time, but existence continues according to Buddhist belief, karmic time is the concern of pilgrims. The mythological narratives about Maratika deal with soteriological matters, mentioning Maratika as a place that can free you from rebirth in the three lower realms: rebirth as animal, as a hungry ghost, or in hell. One young female pilgrim explained to me how she considered it "easier for the gods to give one a better rebirth" after a visit to Maratika than if one had not made a visit. Climbing through the Four Secret Paths (*gsang lam bzhi*) indicates the karmic situation of the pilgrim and a passage will cleanse from accumulated bad karma (Ngawang Chöphel 2000b: 36–7). When the karmic balance of pilgrims are spatialized as their body and the Buddhist karmic law spatialized in landscape the two can be juxtaposed. This tactile use of landscape in rituals brings pilgrims in close contact to landscape in order to help them delimit bad karma, beneficial in this human life, but ever more so in a belief about rebirth according to their karmic time. The passages are referred to as intermediate states (*bar do*), as the transition between death and rebirth, and a passage between stalagmites anticipates the passage between reincarnations. One of the passages in Maratika is even said to indicate how easy and peaceful or difficult and frightening the intermediate state after death will be. The cultural model of karmic retribution and rebirth materialized in rocks and dripstone formations enables the pilgrims to perceptually and experientially anticipate their own death and rebirth.

Conclusion: Spatial Dimension of Times

Material anchors give the possibility to experience by the perceptual system what otherwise could only be conceptualized mentally or in immaterial, cultural models. Therefore, pilgrims do not need strong mental representations of mythology when visiting Maratika, instead they can go and experience, an experience that is choreographed by the layout of paths and passages of the landscape and the tradition from earlier pilgrims. Through ritual practice, the landscape is slowly modified—paths are trodden and geological significant features made more significant by being polished through repeated touch. This makes landscape responsive to mythology through ritual practice, but likewise sustains the mythology. The landscape of Maratika is not only spatialized mythology, like a watch spatializes time; the mythology also temporalizes landscape, because Padmasambhava's initial acts left traces in the landscape, traces that are now blessed. As described, in the landscape, self-produced manifestations are believed to emerge. Indeed, landscape changes through slow geological processes, due to weather conditions (during the last decade, a landslide has "opened" one cave in Maratika thus changing it into a gorge, just as new caves have been exposed) and

Figure 8.4 Climbing through this passage will, according to tradition, indicate the pilgrim's intermediate state between death and rebirth

through ritual practices as just mentioned. These changes further strengthen the belief of an ever unfolding mythology taking more and more space in Maratika.

In Maratika, the pilgrimage site is established as a sacred site by the alleged mythological activities of religious agents in the past, upheld by religious specialists' re-telling and elaboration of these mythological activities, and the pilgrims' and visitors' consumption of the products of these activities through an interaction with the landscape. Maratika is thus situated between semantic construction, narrative elaboration and perceptual interaction.

Through the processes of temporalization and spatialization, thinking about religious themes such as longevity and immortality can be extended in ways allowing the geographical landscape to be understood as part of the evolving mythology. Mythology, dealing with the aforementioned themes, can be approached perceptually through the landscape. Spatialization gives mythological events a material manifestation in space, and mythology then taken to be perceptually available for believers. When pilgrims walk in the landscape, they also move in mythology; when they collect water and pebbles, they also collect blessings from Padmasambhava; when they climb through narrow passages in rocks, they also test their karmic balance and cleanse karmic defilement—all practices that in different ways relate to long human life time and a favorable rebirth in karmic time.

In addition, by temporalization, I refer to the way in which landscape is given a temporal dimension through its inclusion in mythology; thus landscape changes and is dynamic because it is part of mythological developments. In the sacred landscape of Maratika a conflation of dichotomies between myth and landscape, time and space, and immortality and longevity takes place. Therefore, in the physical landscape of Maratika, pilgrims are able to experience three different times in one place and to relate these times to each other. Specifically, the landscape of Maratika is a materialization of the teaching of immortality and thus blessed with the qualities of longevity appreciated by pilgrims. This is achieved through narrative. The guidebook describes the landscape and its geological features and conflates with mythology. Pilgrims, dealing with karmic residues, can therefore negotiate time, when they perceptually interact with the material landscape—a landscape possessing Padmasambhava's blessing and life-enhancing qualities.

References

Aris, M. (1988) *Hidden Treasures and Secret Lives*. London: Kegan Paul.

Bender, B. (2002) Time and landscape. *Current Anthropology*, 43(S4): 103–12.

Brandt, P.A. (2005) Mental spaces and cognitive semantics: a critical comment. *Journal of Pragmatics*, 37(10): 1578–94.

Buffetrille, K. (1994) *The Halase-Maratika Caves (Eastern Nepal)*. Pondicherry: Inst. Français.

Buffetrille, K. (1998) Reflections on pilgrimages to sacred mountains, lakes and caves. In A. McKay (ed.), *Pilgrimage in Tibet*. Richmond: Curzon, pp. 18–34.

Corrigan, J. (2008) Spatiality and Religion. In B. Warf and S. Arias (eds), *The Spatial Turn: Interdisciplinary Perspectives*. Abingdon: Routledge, pp. 157–72.

Day, M. (2004) Religion, off-line cognition and the extended mind. *Journal of Cognition and Culture*, 4(1): 101–21.

Doctor, A. (2005) *Tibetan Treasure Literature: Revelation, Tradition, and Accomplishment in Visionary Buddhism*. New York: Snow Lion Publications.

Garrett, F. (2009) The alchemy of accomplishing medicine (sman sgrub): situating the yuthok heart essence (G. yu thog snying thig) in literature and history. *Journal of Indian Philosophy*, 37(3): 207–30.

Gayley, H. (2007) Soteriology of the senses in Tibetan buddhism. *Numen*, 54(4): 459–99.

Grapard, A.G. (1982) Flying mountains and walkers of emptiness: toward a definition of sacred space in Japanese religions. *History of Religions*, 21(3): 195–221.

Gyatso, J. (1996) Drawn from the Tibetan treasury: the gTer ma literature. In R.R. Jackson, J.I. Cabezón and G. Lhundrup Sopa (eds), *Tibetan Literature: Studies in Genre: Essays in Honor of Geshe Lhundrup Sopa*. New York: Snow Lion, pp. 147–69.

Huber, T. (1999a) *The Cult of Pure Crystal Mountain: Popular Pilgrimage and Visionary Landscape in Southeast Tibet*. Oxford and New York: Oxford University Press.

Huber, T. (1999b) Putting the gnas back into gnas-skor: rethinking Tibetan pilgrimage practice. In T. Huber (ed.), *Sacred Spaces and Powerful Places in Tibetan Culture: A Collection of Essays*. Dharamsala: Library of Tibetan Works and Archives, pp. 77–104.

Huber, T. (2008) *The Holy Land Reborn: Pilgrimage & the Tibetan Reinvention of Buddhist India*. Chicago and London: University of Chicago Press.

Hutchins, E. (2005) Material anchors for conceptual blends. *Journal of Pragmatics*, 37(10): 1555–77.

Kapstein, M. (2000) *The Tibetan Assimilation of Buddhism: Conversion, Contestation, and Memory*. Oxford and New York: Oxford University Press.

Karma Wangchuk (ed.) (2000) *The Guide to Maratika*. Kathmandu: Maratika Chimey Takten Choling Monastery.

McCutcheon, R.T. (2000) Myth. In *Guide to the Study of Religion*. London and New York: Continuum, p. 208.

Namkhai Norbu (1992) The pilgrimage to Maratika. In N. Norbu (ed.), *Dream Ioga and the Practice of Natural Light*. Dharamsala: Snow Lion Publications, pp. 73–89.

Ngawang Chöphel (2000a) Biography of Maratika Lama. In K. Wangchuk (ed.), *The Guide to Maratika*. Kathmandu: Maratika Chimey Takten Choling Monastery, pp. 54–74 (in Tibetan) and pp. 40–54 (in English).

Ngawang Chöphel (2000b) Introduction to Maratika. In K. Wangchuk (ed.), *The Guide to Maratika*. Kathmandu: Maratika Chimey Takten Choling Monastery, pp. 35–53 (in Tibetan) and pp. 25–39 (in English).

Østergaard, J. (2010) A topographic event: a Buddhist lama's perception of a pilgrimage cave. *Social Analysis*, 54(3): 64–75.

Østergaard, J. (2011) *The Topographic Mind: Connecting Cognition to Landscape in the Study of Religion – with Empirical Illustrations from Maratika, a Pilgrimage Site in Nepal*. Unpublished, Aarhus University.

Pordié, L. (2008) Reformulating ingredients: outlines of a contemporary ritual for the consecration of medicines in Ladakh. In M. V. Beek and F. Pirie (eds), *Modern Ladakh: Anthropological Perspectives in Continuity and Change*. Leiden and Boston: Brill, pp. 153–74.

Robbins, P. and Aydede, M. (2009) *The Cambridge Handbook of Situated Cognition*. Cambridge and New York: Cambridge University Press.

Samuel, G. (1993) *Civilized Shamans: Buddhism in Tibetan Societies*. Washington, DC and London: Smithsonian Institutions Press.

Strong, J. (2004) *Relics of the Buddha*. Princeton: Princeton University Press.

Thondup, T. (1997) *Hidden Teachings of Tibet*. Somerville: Wisdom Publications.

Chapter 9

Creating the New Times: Reburials after War in Northern Uganda

Lotte Meinert and Susan Reynolds Whyte

This chapter explores the creation of temporal beginnings and continuities through reburials in Northern Uganda. After 22 years of armed conflict, the Acholi are moving from Internally Displaced People's camps back to their rural villages. In the process, they must deal with bones that are out of place in order to usher in a new time. During the years of conflict, survivors have buried most of their dead within the camps on land that does not belong to their own clans. In Acholi cosmology this constitutes serious disorder as the dead are supposed to return to patrilineal clan land to join the living family as well as ancestral spirits. The dead are being exhumed and reburied in their proper homes, not only to create spatial order, but also, we argue, to create and re-create temporal order. The bones are associated with the spirits (*tipu*) of the dead, who are said to be dissatisfied if not taken home. Meanwhile, as people begin to clear bush to start farming again, they sometimes find the bones of unknown fighters killed in the war. These remains need special rituals in order to avert the dangerous spirits (*cen*) of those who died badly with no one to take them home for burial.

We explore these processes and rituals of exhumation, reburial and cleansing as social technologies of time manipulation. As in other aspects of Acholi cosmology, sequence and timing are fundamental. In their disposition of bones, people are attempting to achieve durability and permanence, on the one hand, and to achieve endings to disorder and beginnings of new, peaceful, times on the other.

Since Robert Hertz's famous essay on death, first published in 1907, scholars have attended to the relation between time, the decomposition of the material corpse, and the regeneration of society and social values, sometimes enacted by dealing with the bones. In the specific historical context of Northern Uganda, where so many people died from violence and disease and were buried 'wrongly', this relationship has taken on special and urgent significance. The reburial of the bones in the ancestral land creates material continuity between land and people, as well as between the living and the dead, which makes new life possible, and at the same time confirms a family's claim on the land. The exhumation and re-interment of bones in their proper homes are in this sense a specific technology of time. They mark not only the transition to the status of ancestor, but also the re-establishment of a home after years in the camp, and a future on the land for the coming generations. Reburial of bones thus links different temporalities in past, present and future: the final end of a living relative, the beginning of a new life on ancestral land, and the continuity of kinship values and entrustment of land into the future.

Rather than giving a generalised description of reburial practices in Northern Uganda, we proceed by presenting a specific case in detail in order to excavate the particularities of the timework put in place in this specific situation. This is followed by a contextualised analysis and conclusion regarding the grasping of time through the materialities of death.

Bringing the Dead Home

It is early morning in a neighbourhood in Gulu town in Northern Uganda. A group of men from Otto Samuel's family is gathered behind one of the round huts at the place where Samuel's brother, Michael Opoka, has been buried since 1995. Michael, a government soldier, was shot by the rebels near Sir Samuel Baker's Fort when he was 35 years old. The army took his body to Gulu town for burial because the war prevented burial at his rural home.

An opportunity has now come and it is time to bring Michael's remains back to the family home. The men have brought hoes for digging, a goat and a knife for making a sacrifice, and two sacks for carrying the remains of Michael Opoka and the sacrificed goat. They start digging and reach a cavity that had contained the coffin, which has now rotted away. More men have turned up at the gravesite. They are all in their fifties or sixties. One man has come because he was the parish chief who buried Michael Opoka 16 years ago.

A piece of blue blanket is uncovered in the grave, and the men start digging carefully with their hands. When the entire body is free from soil, four men lift the remains out of the grave. There is little left inside the blue blanket but the bones and the skull. The men put them, unceremoniously, in a white sack, feet first, and stuff in all the bits of coffin they have dug up too. The bag, now containing the remains of Opoka, is put aside, up against the hut, and they all turn to the next job. A small boy is standing by the grave holding a goat on a rope. It is nibbling on some grass, oblivious of what is to come.

One of the diggers drags the goat closer to the grave and, with the help of Opoka's son, makes the goat lie down and cuts its throat with a small blunt knife. The pulsing of red blood on the man's dusty hands and on the ground makes a sharp colour contrast. The man opens the stomach of the goat, takes a bit of green chyme from the intestine, and sprinkles it thrice into the grave snapping his fingers. His movements are quick and professional; there is very little solemnity about the situation and he does not say anything. The bones of Opoka in the white sack and the slaughtered goat in the other are placed in the back of the car. Opoka's son and one of the diggers come along in the car with the bones and the goat. They are not supposed to stop on the way, but they make a halt at the market to buy a white sheet for wrapping the bones to be placed in the new grave. They drive in near silence the 47 km to Samuel's home in Onguti village.

Figure 9.1 Excavating the coffin (photograph by Sophie Hooge Seebach)

A large group of people is congregated in the home. Men are sitting on benches, and a group is standing around the new grave, among them Otto Samuel himself. The hole has already been dug, a large mound of red earth on each side of it. A pile of red bricks has also been placed next to the grave. The women of the family sit on a mat in the shade of a small tree, close to the new grave. Among them is Samuel's elderly mother, born in 1911. The sack with Opoka's bones is brought to a large papyrus mat where the bones are rearranged and folded into the new white sheet. The skull rolls out, but is put back into the package together with the pieces from the old coffin. The bones are wrapped as a body and carefully lifted into the grave. The soil and pieces of the old coffin left on the papyrus mat are poured into the grave and the men start filling the grave with soil. They finish the new grave with a neat layer of bricks. Meanwhile the goat's meat is being prepared for a meal. Samuel says that they are now relieved and it is time for the family to eat and drink and celebrate that their family member, Michael Opoka, has finally come home.[1]

1 This reburial was observed and video filmed by Sophie Hooge Seebach, who is a student doing fieldwork in the area on issues related to death; she was also a field assistant on the project at the time. We draw upon her field notes from this reburial as well as our notes from several interviews and visits with the family.

Figure 9.2 Taking the sack of bones, followed by the sack with the sacrificed goat, to the vehicle (photograph by Sophie Hooge Seebach)

Figure 9.3 Rearranging the bones and skull in a new sheet before reburial in the home (photograph by Sophie Hooge Seebach)

Figure 9.4 The remains have been reorganised as a body ready for replacement in ancestral soil (photograph by Sophie Hooge Seebach)

Figure 9.5 Marking the grave precisely with bricks as a lasting sign that will be visible for future generations (photograph by Sophie Hooge Seebach)

Otto Samuel is 69 years old, a farmer and carpenter who lives in Onguti village in Awach sub-county with his old mother, his brothers, his children and his sister. They moved back to this home after the Internally Displaced People's camp in Awach was closed in 2008. In this family, as in many others, the older men went back to the original homes first, followed by the women and children. The family members who had died during the conflict and were buried in the camp remained there for three more years and were brought home in 2011. In addition to Michael Opoka, Otto Samuel's family reburied a foetus and three men – a father's brother, a brother and a sister's son – who had been left in the camp.

Otto Samuel explained that some time after they moved back to the village, a bicycle fell and broke his granddaughter's leg, which did not heal well. Some of the family members found the incident disturbing and went to an *ajwaka* (diviner) to enquire. The *ajwaka* divined the case and found that it was the old *Mzee* (an elder from their family, a father's brother to Samuel) buried in Awach who was complaining: 'It is me, you should take me back.' The family organised to exhume him from the ex-camp in Awach. They sacrificed a male goat and made a new grave for *Mzee* at a proper distance from the house. On top they planted a tamarind tree, because he had died from leprosy; this tree, said Samuel, is a sign that will make the future generations know that this grandfather was a leper. After the reburial there had been no more accidents in the family, said Samuel.

On the same day two more men and a male foetus were reburied; one sacrificed goat can serve for several exhumations as long as it is of the same sex as the deceased, so there was an economy of scale in dealing with all the males at once. One of them was Samuel's brother, who was 54 years old when he died from illness in the camp. He was reburied in the village home, on the opposite side of the houses from the leprosy grave. Behind his grave, towards the grandmother's house, was another grave, barely visible for the sweet potatoes growing around and on top of it. Samuel spoke very quietly about this grave, as if to avoid being overheard. It was the grave of his sister's son who died when he was 27 years old. He had been a rebel leader of high rank, but had deserted the rebel army during the conflict. One day a government soldier shot him when he was riding on a *boda*, a bicycle-taxi, in Awach camp. Samuel said that the soldier must have recognised him from when he was a rebel leader, but he had become a civilian and there was an amnesty so they were not supposed to shoot him. Samuel explained that the family was planning to ask for compensation from the government, because the army killed their nephew unlawfully, but Samuel doubted the government would pay it. The nephew had lost his parents early in life, so Samuel had been taking care of him and that is why he was reburied at Samuel's home. A male foetus, which had first been buried in Awach IDP camp was also exhumed and reburied in the home, at the veranda of the house of its parents.

Burials and the Temporality of Corpses

The specific historical context of Northern Uganda, where hundreds of thousands of people died from violence and disease and were buried 'out of place' in Internally Displaced People's camps, adds a new facet to the classic anthropological research on burials and death. We will argue that the reburials of these displaced remains may be seen as a specific kind of timework. Like all steps in the cycle of funerary rituals, they deal with the process of individual dissolution and social continuity. They also refigure history by creating a new epoch and attempting to elide a period of terrible disruption.

In his seminal essay on death, Robert Hertz urged us to pay attention to what happens to the corpse over time, and he was a pioneer in theorising how the temporal decomposition of the material corpse relates to conceptions of the soul and the afterworld, that is, how the material and the spiritual are two sides of the same dead body. The significance of the corpse in burial rituals is that in dealing with the material remains people can deal with the spiritual as well. By acting on the tangible and visible corpse and its decomposition, they make the intangibility and invisibility of the soul and its transformation more comprehensible. Hertz wrote:

> there is a close relationship between the representations of the body and that of the soul. This mental connection is necessary, not only because collective thought is primarily concrete and incapable of conceiving a purely spiritual existence, but above all because it has a profoundly stimulating and dramatic character … The material on which the collective activity will act after the death, and which will be the object of the rites, is naturally the very body of the deceased. The integration of the deceased into the invisible society will not be effected unless his material remains are united with that of his forefathers. It is the action of society on the body that gives full reality to the imagined drama of the soul. (1960: 83)

Taking our cue from Hertz we focus our attention on the destiny of the corpse, but we expand on the temporal aspects of burial rituals. We argue that in the reburials in Northern Uganda people deal not only with the intangibility of the spiritual, but also with the intangibility of the temporal. Transience, permanence, durability and finality are difficult phenomena to grasp, but made evident in death, burials and not least in reburials. We suggest seeing the re-placement of bones and graves in Northern Uganda as a kind of timework through which people deal with and appropriate different temporalities. We are interested in the concrete activities and necessary materials, but also in the procedures, which involve ordered steps, duration, frequency, sequence and timing.

We borrow the notion of timework from Michael Flaherty who describes timework as 'intra-personal and inter-personal effort directed toward provoking or preventing various temporal experiences' (2011: 11). Whereas Flaherty is interested in self-selected agentic micro-management of temporal experience (inspired by Goffman's interest in Face-Work), our attention is primarily directed towards the socialities and collective frames of timework involved in reburials. The re-placement of bones and graves in the geographical and social landscape, the ordering of hierarchies and sequences by placing bones in patterns of proximity, distance and orientation, are ways of understanding and doing time through place and materials.

British anthropologists who studied funeral practices were heavily influenced by Hertz. They tended to take the Durkheimian strand in Hertz's work, that is, the centrality of the social, though unlike Hertz, they did not see society as an actor: 'it is not so much a question of Hertz's reified "society" responding to the "sacrilege" of death, as of the mortuary rituals themselves being an occasion for *creating* that "society" as an apparently external force' (Bloch and Parry 1982: 6). Goody's (1962) formidable study of death rituals and intergenerational transmission of property rights among the LoDagaa of West Africa, and Bloch's monograph on burial rituals of the Merina of Madagascar, were about the re-creation of social forms and values through ritual management of death. These themes were sharpened in the collection entitled *Death and the Regeneration of Life* (Bloch and Parry 1982) in which the editors argue that death is the beginning of a ritual process of regeneration, overcoming individual biological mortality by reasserting lasting fundamental social values. In a recent edited volume on funerals in Africa, Jindra and Noret remind us that burials are often key cultural events in which people invest large sums of money and effort so as to conduct them in 'the right way' – and what is considered 'right' undergoes remarkable historical change. Importantly, they also point out, in contrast to the older Durkheimian-inspired analyses of funerals, that burials are not only moments of social re-integration in the face of death, but are occasions for both (re)production and (un)making of solidarities and hierarchies, alliances and conflict (Jindra and Noret 2011: 2) We acknowledge our indebtedness to these social anthropological studies of funeral rituals, but we need other analytical tools as well to do justice to the conditions in which a whole generation of people like Otto Samuel are trying to get on with their lives and create a new epoch in history.

What we suggest here is that the reburials in Acholiland accomplish timework in two different scales simultaneously. On the one hand, they fit into the cycle of funeral ceremonies that deal with the temporal processes of material decomposition, spiritual transformation and social regeneration. They ensure the continuity of the shades in the community of the living on a personal familial scale, in the ways that Bloch and Parry (1982) highlight. On the other hand, these reburials are occurring in a particular historical and political context, and they work upon a broader scale to create a new epoch in Acholi history.

In her book on reburial and postsocialist change, Verdery argues that the materiality of dead bodies makes them potent means of reconfiguring time: 'Bodies have the advantage of concreteness that nonetheless transcends time, making past immediately present' (Verdery 1999: 27). She acknowledges the personal and familial temporality of a dead body, which recollects a life with others, processes of growth and aging, the punctuation of death, the moral imperative of proper burial, the continuing respect for the memory or soul of the deceased. But she is most concerned with the way this scale of temporality is conflated with and vitalises the politics of historical transformation. Under conditions of radical political change, reburials can work upon the very understanding of temporal process and the shape of history (Verdery 1999: 115). The sense of history takes on an aura of truth and reality made tangible in the bones: 'Manipulating physical remains is a visual and visceral experience that seems to offer true access to the past' (Verdery 1999: 113). Reburials on a large scale allow people to excise a portion of the past, to bracket it as an aberration that can be cut out, so that they can 'repossess a "normal" past and weave it into "normal" presents and futures' (Verdery 1999: 116). Moreover, they give another shape to history by enacting notions of genealogies that are inclusive, but also exclusive. Compared to the broad community of the camps, the kin groups who mobilise to rebury their dead are narrow; but their depth is greater than the temporal frame of war and encampment. Reburials place the dead in a context of ancestors and descendants, with a past and a future that were amputated in the camps.

Return Time and the Remaining Remains

The war between the insurgent Lord's Resistance Army (LRA) and the national Uganda Peoples' Defence Force (UPDF) was fought most bitterly in Acholiland. Although the fighting spread east to where the Langi and Teso people lived, and north to the Sudan where the LRA had important bases, it was the Acholi region that suffered longest. Beginning in the 1990s the UPDF adopted the strategy of forcing the civilian population into Internally Displaced People's camps established around trading centres and schools. By 2006, when the Cessation of Hostilities Agreement was signed in Juba, about 95 per cent of the population, nearly two million people, were living in these congested conglomerations of thatched huts. In principle, the camps were for the protection of civilians, and each was defended by a detach of UPDF soldiers. Camp perimeters were sharply defined and 'for their own safety' people were subjected to strict curfews; they were only allowed to cultivate and collect firewood within a few kilometres of the camp, near the roads that were patrolled by the UPDF (Dolan 2009: 119). In practice, the UPDF was unable to protect the camp residents from abduction and killing; not only were the soldiers ineffective, they were widely feared for their harsh methods. The government forces suspected locals of aiding the enemy, since the LRA consisted almost entirely of Acholi; some had joined voluntarily and many had been forcibly impressed. The LRA war was a civil war with Acholi families involved on both sides, and, like the family of Samuel Otto, suffering losses to both armies.

For up to two decades in some areas, people were confined, not to some safe place away from the fighting, but within the war zone, caught between the LRA and the UPDF, close to their homes, but forbidden to return to them. During lulls in the fighting, people moved to other camps to join relatives; convoys allowed some movement along the main roads. But for most Acholi, the years of the war were years of immobility and dependence on food rations from the World Food Programme. The deprivation, humiliation and uncertainty of those years have been described by researchers (Finnström 2008; Dolan 2009; Branch 2011), but the problem of burying the dead has hardly been mentioned. Unable to return to their homes to place their dead where they belonged, people had to dig graves in the camps. Those who died violently, children who succumbed to malaria or malnutrition, adults who surrendered to AIDS, aborted foetuses, stillbirths and old people whose lives ended in the camps: all had to be interred in soil that belonged to someone else. While some of the IDP huts were put up on land belonging to churches, schools, health units or government, most were on the property of families who happened to live in the areas gazetted as camps. There were no designated cemeteries in the camps. The landowners had to accept, not only the huts that were erected on their land, but the growing number of graves between them. Women were buried behind the house near the fireplace, men in the front, miscarriages and stillbirths next to the wall of the hut. Graves were marked by stones and bricks, occasionally by a cemented tombstone.

From 2006, the camp policy changed. Greater security prevailed and the authorities opened 'satellite camps' or 'transit sites' so that people could move out of the crowded main camps to locations closer to their land. The idea was that they could start clearing, cultivating and building houses on their 'ancestral land' while enjoying the relative safety of the satellite camps. Return time had its own rhythm. People waited for the dry season when thatching grass was easily available. Often some members of the family moved back first, leaving children in the camps close to their schools. Those without strength or resources to build their own huts had to wait. In many areas, water supplies, schools and health facilities had to be re-established before the whole family could return. People hesitated because they feared that security was not assured; Joseph Kony, the LRA leader, was still alive. They were also worried about *cen*, the spirits of the unknown dead, whose remains were sometimes found on the land to which they were returning. The cleansing of rural areas in order to make them fit for farming and building was part of the rhythm of return for some villagers.

Acholi people explain that when someone dies, the *tipo*, the shadow or shade, remains near the corpse. Working on the dead body properly involves burying it at home and calling the shade to come home (*omo tipo pa lato*). Exactly where the shade stays, whether inside the house, near the grave, in trees, or in the *abila* shrine, is vague. One woman dismissed our too literal interest in location: 'How can you know where the *tipo* stays—it is not seen.' The important thing is that it protects the home, and that it is associated with the corpse. Labol Jenipher, a diviner, explained that in life spirits are within us, 'like if we dream'. 'But when I die the *tipo* keeps on checking: "Is the flesh there? Are the bones there?"' People described to us how corpses should be buried on their sides, facing into the compound. In the same way, the shades should be oriented to the home and its well-being.

The dark inverse of the well-disposed shade is the spirit of a person who died badly and was never brought home to be buried. The *cen* is annoyed and resentful; it could be someone who was murdered or who died after being poorly treated. Spirits of the vengeful dead that afflict the living are known in many parts of East Africa; but in Acholiland they are particularly associated with unclaimed, unburied bones in the bush. In Paibona, we heard of returnees coming across skeletons with gum boots or tattered bits of clothing suggesting they had been LRA fighters. People explained that the LRA left their dead where they fell; several whose abducted family members had died in the bush recounted sadly that their bodies must be rotting somewhere in the hills. Inadvertently stumbling on such remains can cause misfortune, often in the form of strange behaviour, possession or madness. The link between the material remains and the bitter spirit of the dead was emphasised by the required response to such a discovery: place a leaf on the bones and say 'I bury you' and perhaps also, 'I am not the one who killed you' (Finnström 2008: 88). It was those who returned to the villages first who had experiences of *cen*. When bones were found, or sometimes even when they were only suspected to be in the area, a cleansing ceremony (*moyo kabedo*) was organised on a community basis. The required sheep or goats were subsidised by USAID in some cases as part of donor support to the return process.

Far more common, however, are the continuing efforts to deal with the known dead who were buried and whose bones and shades should return to their villages, just as their living relatives are doing. This work is a family matter; it is the close relatives of the deceased who must acquire the goat to be sacrificed and organise the people to carry out the tasks. Very many families have left graves behind in the former camps. It is not so much that the shades of these dead are vengeful and frightening like the *cen* in the bush – though misfortunes can be attributed to their displeasure as we saw in the case of Samuel Otto. Rather there is a sense that return is not finished until the bones and *tipo* of the dead come home. *Cen* are spirits of those who have been wronged, whose bodies were abandoned. They are the negative image of the moral imperative to care for the material and immaterial remains of the dead. Exhuming the remains and reburying them at home shows that you respect and love your relatives – that you have not neglected or abandoned them.

The timing of exhumation and reburial varies according to family finances, misfortunes attributed to the shades, and pressures exerted by the owners of the land in the IDP camp. Some brought the dead home soon after they themselves returned; one man told how, pressed by the landowner in the camp, he brought four corpses in one day, each rolled in two sheets and a papyrus mat and carried on the back of a bicycle. Others had reburied only some of their dead relatives and had definite plans for the return of the remaining remains. Many asserted that they wanted to move the dead, but lacked the means to do so and thus could not set a definite time. One man, still living in the former IDP camp because he had not yet completed houses in the village, recounted that all of his dead relatives had been exhumed. Now he was surrounded by their open empty graves and longed for the day when he would join the whole family, living and dead, in their ancestral village.

For those who were returning, moving the remains of the dead was part of the process: 'Otherwise you feel that you have left something undone.' It was also important for those upon whose land the dead of other families had been interred. Landowners in Awach, a former IDP camp, could count the number of graves remaining on their land. One woman told how she had been round to visit the village homes of those who had stayed on her land to request them to come back for their dead. Although many landowners were sympathetic, they found it awkward to host the bones of the displaced dead. The reasons were several. One schoolteacher explained that his wife was afraid to dig near the graves for fear of the shades that might be hanging about. To drive a plough or tractor over a grave would be disrespectful. Nor could one build on the spot. Potential buyers would not want to purchase land with graves. The shades might be dissatisfied and disturb the landowners. Most of all, the dead should be with their own relatives. Only in cases where the deceased were of the same clan as the landowner, did both landowner and bereaved family agree that the remains might remain where they were.

Ritual Chronology: The Work of Burying and Reburying

Reburials can be seen in the context of normal burials. Our informants, and other researchers (Odoki 1997), describe the sequence of funeral rituals in Acholiland: first there is the ritual of burying the corpse (*iko liel*) with its necessary steps including the calling home of the shade; then the second funeral rite of 'smearing the grave' (*puyo liel*) after three or four days, which confirms the death and the place of the *tipu* and cleanses the mourners; and finally the last funeral rites of memorial, which might be delayed up to 10 years or postponed indefinitely, depending on the status of the deceased and the wherewithal of the family. These events establish a characteristic temporal order and rhythm in death and interment, which resemble the process that Hertz (1960) describes so insightfully, and establish a connection between soil time and soul time as the flesh decomposes in the earth. From the moment of death, a state of liminality prevails that only ends provisionally with the smearing of the grave, which cleanses the grave diggers and their tools, and permits normal life to recommence.

Unlike the Dayak people of Borneo, about whom Hertz wrote, or the Merino of Madagascar (Bloch 1971), Acholi did not practice reburial as part of their ordinary funeral cycles, until it became common practice because of the war. However, there was and is still an important exception. The children of Jok (*latino jok*) are special people – twins, children who are born in breech position, with three nipples, an extra set of little fingers or another unusual corporeal mark. They possess or are connected to special powers,[2] and are believed to have *hot spirits*. For that reason they always need to be reburied. After a normal first burial, they are left in the soil for about a year until only the bones remain. After this period of soil time, the children of Jok are exhumed and their bones are put in pots with holes for ventilation and vision, and the pots are placed in the middle of the compound, where they can 'keep an eye' on the home. Other corpses, besides those of Jok's children, were occasionally moved, as well. Thus, the practice of exhumation and re-interment was not unknown to Acholi people, even before the current wave of reburials. Perhaps the post-conflict reburials might count as invention of tradition (Hobsbawn and Ranger 1983), but they also resonate with older continuing practices. In any case, some people already had experience handling bones of the dead.

The ongoing reburials in Acholiland are so clearly about creating order in space: the dead lying in strangers' land are 'matter out of place', which constitute potential disturbance as the dead are supposed to be buried in patrilineal clan land. But it is not enough to bury them on clan land. When they are re-placed in families' original homesteads, they re-create social orders and hierarchies. The spatial patterns of placing graves vary among clans, but most put children and foetuses close to the house, women near the kitchen and men closer to the road. Graves are usually oriented east-west and some Acholi explain that this is parallel to the path of the sun from sunrise to sunset. Special people, such as twins, or those who died special deaths, such as from leprosy, are placed differently (Odoki 1997). These patterns were followed in reburials too, as we saw in the case of Otto Samuel's family members. (Another family told us that all their reburied dead were placed facing inward to the compound, except those of a murdered man whose face was oriented outwards towards the bush, so his shade could go looking for his murderer.) The ordering of space with the graves in the homesteads is material and permanently visible. But the *ordering of time* is chronologically tangible only

2 The power called Jok in Nilotic languages has been discussed by many scholars, and is reviewed by p'Bitek (1971). Mogensen writes: 'The Nilotic notion of jok has been used to emphasize that complex cosmological and philosophical systems did exist in pre-colonial Africa. While missionaries searched for ways to translate "God" and communicate with people about Christianity, ethnographers (and others) have sought logic and coherence through interpretations of juok in terms of universal force or "power"' (Mogensen 2002).

in the process of exhuming and reburying. Afterwards several temporalities are conflated, as we suggest below.

Everyone we have talked to affirmed that eventually they want to bring their dead home (unless they are buried on the land of a clansman). However, as long as the graves, the remains and *tipo* of the dead are not disturbed or disturbing, people can bide their time, waiting to assemble the necessary resources. Other concerns may be more urgent. What triggers the initiation of reburial varies: an opportunity arises; the landowner makes trouble; or misfortune strikes among living relatives – as in Otto Samuel's case. This changes the urgency and timing perspective: a ritual chronology starts which is distinct from everyday rhythms and occurrences. If ritual time is set in motion by a misfortune, the first move often includes a visit to a diviner, who will identify causes and suggest measures to alleviate the affliction. The diviner might find that problems are caused by dead shades longing to go home to be among and honoured by the living, like Samuel's uncle who communicated, 'It is me, you should take me back'.

The order in which the dead are brought home differs among families, but there is a general consensus that order is significant. Many families specify that the order must follow the chronology of deaths: the person who died first has to be brought back first, followed by the one who died next, etc. Age, gender and status make no difference; a foetus can be returned before an elder. In this way the reburials recollect the time lived during the war as a material death calendar. But once interment is complete, of course, they are all simultaneously present in the home. In other families, however, it appears that status and male gender have been given chronological priority.

When sequence is determined, relatives start searching for the ritual resources for doing the reburial: a goat for sacrificing, money for transport of the body, alcohol, food for the celebration. Announcements have to be made to relatives travelling from afar to participate in the ritual. The preparation time for organising the ritual is often stressful and there is a risk that things get stuck because people cannot afford the sacrifice or the transport or something interferes with the process. When matters go right and all elements are finally ready, there should be no hesitation, the ritual must happen. The *tipo* of the dead will get impatient if people delay the ritual for no reason. There is a sense of urgency and attention to rules and sequences that must be followed. The ceremony is organised in sequences that vary in length and intensity: digging for the bones, exhuming the bones, sacrificing the goat, transporting bones and sacrifice, burying in the new grave, celebration and cleansing of the vehicle utilised. Some sequences take a long time but carry little significance (such as the digging) other sequences are short but semantically intense (such as the ritual killing). The ritual intensifies when the remains in the grave appear and a sense of procedural necessity persists until the bones are back in the soil. The procedural ordering during this phase of the ritual is emphasised by repetition and differentiation: the chyme from the goat is sprinkled thrice in the grave for a man and four times for a woman, a pattern that is repeated in the new grave, but with soil.

In terms of ritual time in this rite of passage, the transport of the remains is a phase of liminality, as Van Gennep (1960 [1909]) would have put it. The risk during this phase of being betwixt and between is that the *tipo* which hangs around the bones has been exposed and might get confused and misunderstand the situation, go astray and transform into an evil spirit, or cause trouble. Special rules have to be followed during the transport; the driver should not greet anyone, look back or stop. Reaching the home, care will be taken to release and cleanse the vehicle that brought the bones, a procedure identical to that used when a vehicle brings a dead body from the hospital for ordinary burial. A chicken will be taken around the vehicle and given to the driver to ensure that the *tipo* is not confused, and does not follow and attack the vehicle, assuming it was the vehicle that caused the death.

Care is taken to collect all remains of the dead including fingernails and toenails and pieces of the coffin; they must be reburied with the bones in one place, to prevent the *tipo* from following a small piece of remains and to reassure the shade that the relatives want it to stay around the home. Reburials are not considered times for mourning, like normal burials, but are solemn occasions for ritual ordering and celebrations of putting things right, and, as we would like to emphasise, to create temporal order, create the new times.

Walking around the former IDP camp in Awach, we saw that the empty graves had not been refilled. People explained that a grave should be left open after exhumation so it can fill in naturally in its own time, as it is washed by rain and accumulates detritus. It was as if they did not want to hurry the 'work of time'. There was no agreement on why a recently excavated grave should not be closed. Most said, 'that is how we do it'. Some explained that it needed to cool, recalling the notion that shades, especially those of Jok's children, could be hot. Perhaps its openness fits with another practice that people also shrugged off as 'that's just our

Figure 9.6 An exhumed grave in the former Awach IDP camp, left open to fill in slowly over time (photograph by Susan Reynolds Whyte)

tradition'. In normal burials, after the body is washed, it is dressed in clothes that are loose. Zippers, buttons and waist elastics are clipped off and left on top of the grave or thrown in the latrine. 'If you leave the buttons it's like putting a stone on the dead body. If you bury with the buttons, illness may strike the family and you have to dig it up and remove the buttons', explained Santa Ajok, a local diviner. The open graves, like the open clothes, allow the *tipo* to be liberated from the corpse and the underground, in order to attend to the home and the family.

Family Past and Future

Reburying relatives on family land is a kind of timework that encompasses past, present and future by materialising and localising the kinship relations that make a home. By interring the remains in the soil that is claimed as ancestral, people make continuity with the past; they enhance a feeling of security in the present by diminishing worries about things left undone and dissatisfied shades; and they anticipate a future of fruitful family unity in and on the earth of a place. The reburials and the enduring graves thus constitute durations: 'convergences of different temporalities within one rhythmic configuration' (Nielsen 2011: 299).

Dead bodies can reconfigure space as well as time; their very corporeality means they have to be some *where*. Verdery (1999: 27) discusses the way reburials in ex-Yugoslavia were important means of localising claims. As in parts of East Africa, kinship was practised as rooted in particular soils (Verdery 1999: 105). Reburials were ways of assigning values to space, values that asserted kinship as a matter of earth. Graves could be claims; in a similar way, graves are used as evidence in land disputes in Uganda.

The assumption of a primordial link between kin groups and land, indeed the overriding importance of descent groups, can be questioned. Writing of another Nilotic-speaking people, the Luo of Western Kenya, Shipton (2009: 86) discusses the increasing importance of graves and their placement. He notes that Luo people did not make grave markers in the past; if anything they used mobile shrines to commemorate ancestors. Now they make monumental cement tombstones, as prosperous families in Acholiland also do. Shipton relates this 'sacralization' of graves to 'the heightened importance of kinship and descent in the reckoning and defending of land claims under conditions of rural crowding and competition for land. Luo people, and especially men, have made graves into *tools of territoriality, and anchors of being*' (Shipton 2009: 96).

Although land shortage is not nearly as acute in Northern Uganda as it is in Western Kenya, land conflicts are rife as people move back to their villages. We coined the term 'patrilineal fundamentalism' to capture the heightened use of biological descent in excluding 'non-belongers' even from areas where their parents or grandparents had been allowed to settle (Whyte et al. 2012). Lamaro Cecilie, one of the people we interviewed about reburials, remembered that before the war, if a family sold their land and moved elsewhere, it was up to the person who bought their land to get the goats for reburials if they did not want other people's graves on their new land. In other words, those who were moving did not feel it imperative to take their dead, suggesting that the material bones and graves have become more important in the period of return when questions of land and belonging are on the agenda.

Yet it would be misleading to understand the timework of reburials as merely anticipating possible future land claims. What is rooted in the land and what people spoke of again and again was home. And home is family. Reburials are explained as reuniting the family, living and the dead, in the framework of a home. Ojok, an old man, who had brought back the remains of his wife, his son's wife, and two children, explained that when you die, it is only your body that dies; your spirit remains alive and will prefer to be where it can see its home and family every day. If you died away from home, you would not be happy until you were reunited with your family, and would keep bothering people until that happened. Rose related how she had reburied her special child, Odoch, sitting in a pot in which two holes had been pierced so his eyes faced the door of their home: 'It is not an animal—a human being should face home.' Many people told of how corpses are buried on their sides,[3] propped up in the grave by stones to keep them in position, looking towards the centre of the homestead. In our imagination, an Acholi home is surrounded by watching subterranean dead.

Creating a home is a matter of entrustment, that is, of placing something in the care of someone for future return or transmission to someone else (Shipton 2007: 11). That something could be land, but it could also be bones and the associated knowledge of kinfolk. Here too there is a simultaneity of past, present and future, as Otto Samuel made clear in telling us about the pedagogy of graves, which instantiate the past for the future. 'The graves will help me to show the children who was here before them. They should grow up and know that this is their home. The most important reason for having the family buried at home is to show the coming generations who is buried there. What is most important is the history.' He made the same point about the grave for his father's brother, who died of leprosy. Graves should be part of everyday life. Samuel noted that people could sit on the veranda, on the very spot where a baby had been reburied. In the same vein, a family in Southeastern Uganda argued that a woman who had left her husband long ago should yet be buried at his home. 'We want to bury her in our courtyard, so that some day when our children are playing and climbing on her grave, we can tell them: "That is the grave of your grandmother Perusi, the one who gave birth to us"' (Whyte and Whyte 2004: 84; see also Whyte 2005).

The past and present family that is gathered for the edification of the future members is in some ways an inclusive and unified one. As Samuel Otto's example showed, men who had fought on opposite sides in the war are joined in family unity after death. Even his sister's son, whose right to inherit land might have been disputed, was brought home to the family that had cared for him. The practice of

3 However, those who can afford coffins place the dead on their backs.

reburial, the disposition of the material remains, may thus also contribute to the creation of new times by excising the divisions and separations that many families experienced during the war. The ideal of home and family unity is made material as a past, present and future, despite the recognition that it may never be achieved.

Conclusion

The disposition of corpses during and after periods of crisis is a matter of concern in many African societies. In concluding, we may consider the Northern Uganda case in relation to other examples of timework and death in periods of disruption. On the one hand, there are situations of destabilisation, displacement and chronic crisis in which the dead are desacralised, neglected by their families and ignored by the state. On the other, there are attempts to make national history by publicly celebrating and memorialising the dead for political ends.

De Boeck (1998) writes of the banalisation of death in the Democratic Republic of Congo, where 'for an increasing number of people, death occurs outside [the social network], and has therefore lost its capacity for emplacement. In a state which is itself adrift and in which many people are on the move displaced for economic and political reasons … it has become increasingly difficult to position the dead' (1998: 50). These neglected dead are restless and active among the living, haunting and disturbing them, and refusing to respect the divide between life and death. 'In the end, in the postcolonial *beyond* … we find no pre- or postmortem any longer, no past and no future, no memory and no oblivion, no dead and no living' (1998: 51). Bodies hastily buried by strangers far from home cannot mark time for their bereaved families and communities.

Werbner (1998) discusses the opposite case of remembrance and memorialisation of war dead by the nation-state, noting its emergence in Europe after the First World War. This has taken different turns in independent African countries. Werbner writes of Zimbabwe where the nation's elite independence fighters are glorified in monumental graves, while common soldiers are not memorialised as individuals in any public way. By contrast, Rwanda celebrates a kind of democracy of death after the genocide, as a national project during annual days of mourning. Skulls and bones are exhibited at over 200 massacre sites throughout the country; other remains are interred in mass graves commemorating the genocide. At these sites, most remains are anonymous because it was difficult to reassemble individual bodies. But the bones and the yearly ceremonies perform timework on a national scale, drawing a terrible line between before and after and urging commitment to a future of 'Never Again'.[4]

4 After the 1983–6 battles in Uganda's Luwero region between the insurgent National Resistance Army and government forces, skulls were put on display to commemorate murders of civilians as well as deaths of fighters. But these bones were removed and buried after a few years.

The people of Northern Uganda are dealing with their war-dead in another way. There are no monuments to warriors and few for slaughtered civilians. The state is leaving the corpses and the memories to their families. And because of the protracted period of encampment, the dead include many whose demise is not directly due to the war. In this respect, the situation resembles more closely the one De Boeck describes, where people died and were buried away from home. But unlike the people of Southwestern Congo, the Acholi are trying to re-place the dead and thereby to work upon the past and the future.

The historical situation of post-conflict reburials in Northern Uganda provides a window, not only for understanding the timework carried out in this specific situation, but also for understanding how the materiality of dead bodies – in general – can be used as a technology of time manipulation that works at several different time scales simultaneously. The concreteness of the dead body offers possibilities. Manipulating it, people try not only to make pasts present, as pointed out by Verdery (1999); they also hope to create 'proper' homes and families in the present, and to work towards potential futures.

References

Bloch, Maurice. 1971. *Placing the Dead: Tombs, Ancestral Villages, and Kinship Organization in Madagascar*. London: Seminar Press.

Bloch, Maurice and Jonathan Parry, eds. 1982. *Death and the Regeneration of Life*. Cambridge: Cambridge University Press.

Branch, Adam. 2011. *Displacing Human Rights: War and Intervention in Northern Uganda*. Oxford: Oxford University Press.

De Boeck, Filip. 1998. Beyond the grave: history, memory and death in postcolonial Congo/Zaire. In *Memory and the Postcolony: African Anthropology and the Critique of Power*, edited by R. Werbner. London: Zed Books, 21–57.

Dolan, Chris. 2009. *Social Torture: The Case of Northern Uganda, 1986–2006*. New York: Berghahn Books.

Finnström, Sverker. 2008. *Living with Bad Surroundings: War, History and Everyday Moments in Northern Uganda*. Durham, NC: Duke University Press.

Flaherty, Michael G. 2011. *The Textures of Time: Agency and Temporal Experience*. Philadelphia: Temple University Press.

Goody, Jack. 1962. *Death, Property and the Ancestors: A Study of the Mortuary Customs of the Lodagaa of West Africa*. London: Tavistock.

Hertz, Robert. 1960 (orig. 1907). *Death and the Right Hand*. Translated by R. Needham and C. Needham. London: Cohen and West.

Hobsbawn, Eric and Terence Ranger (eds). 1983. *The Invention of Tradition*. Cambridge: Cambridge University Press.

Jindra, Michael and Joël Noret, eds. 2011. *Funerals in Africa: Explorations of a Social Phenomenon*. New York and Oxford: Berghahn Books.

Mogensen, Hanne O. 2002. The resilience of Juok: confronting suffering in eastern Uganda. *Africa* 72(3), 420–436.

Nielsen, Morten. 2011. Futures within: reversible time and house-building in Maputo, Mozambique. *Anthropological Theory*, 397–423.

Odoki, Sabino Ocan. 1997. *Death Rituals among the Lwos of Uganda: Their Significance for the Theology of Death*. Gulu: Gulu Catholic Press.

p'Bitek, Okot. 1971. *Religion of the Central Luo*. Nairobi: East African Literature Bureau.

Shipton, Parker. 2007. *The Nature of Entrustment: Intimacy, Exchange, and the Sacred in Africa*. New Haven: Yale University Press.

Shipton, Parker. 2009. *Mortgaging the Ancestors: Ideologies of Attachment in Africa*. New Haven: Yale University Press.

Van Gennep, Arnold. 1960 [1909]. *The Rites of Passage*. Translated by M.B. Vizedom and G.L. Caffee. Chicago: University of Chicago Press.

Verdery, Katherine. 1999. *The Political Lives of Dead Bodies: Reburial and Postsocialist Change*. New York: Columbia University Press.

Werbner, Richard. 1998. Smoke from the barrel of a gun: postwars of the dead, memory and reinscription in Zimbabwe. In *Memory and the Postcolony: African Anthropology and the Critique of Power*, edited by R. Werbner. London: Zed Books, 71–102.

Whyte, Susan Reynolds. 2005. Going home? burial and belonging in the era of AIDS. *Africa* 75(2), 154–72.

Whyte, Susan Reynolds and Michael A. Whyte. 2004. Children's children: time and relatedness in Eastern Uganda. *Africa* 74(1), 76–94.

Whyte, Susan Reynolds, Sulayman Babiiha, Rebecca Mukyala and Lotte Meinert. 2012. Remaining internally displaced: missing links to security in northern Uganda. *Journal of Refugee Studies*. doi:10.1093/jrs/fes040

Chapter 10
Memory and Succession in the City of the Dead: Temporality in the Ancient Egyptian Mortuary Cult

Rune Nyord

Introduction

Death plays a crucial role in the ancient Egyptian monumental discourse, as some of the most enduring and iconic Egyptian achievements – pyramids, decorated tombs and mummies – are centred squarely on ancient Egyptian conceptions and experiences of death.

The Egyptian tomb serves, perhaps even more so than other fixed ritual locales, as a kind of lens capturing and diffracting the past, present and future. The past life of the tomb owner is fused in a monumental shape with depictions serving as a matrix for the recurrent cult taking place there, while the tomb owner in turn makes promises for the future wellbeing of those serving in the cult. Perhaps the most striking example of the Egyptian attitude to time in death is found in the practice of mummification where the 'eternally' continued existence of the deceased is made possible by giving the dead body a new, permanent shape.

A direct way to approach these questions of time is through the medium of the tomb inscriptions, providing a glimpse into the mode of existence of the ancestor in the tomb. The most immediately accessible part of the inscriptions is the narrative autobiographical texts providing an overview of the highlights of the tomb owner's career. Other inscriptions relate more directly to the function of the tomb and the social and ritual relations between the tomb owner and the tomb visitor. Each of these parts of the tomb inscriptions exhibits its own relationship to time and temporal processes for which the tomb provides a nexus.

One of the most challenging problems in the understanding of the ancient Egyptian mortuary cult is the particular connections between the gathering of times in the tomb and the notions of the creative potential inherent in the immutable and hidden mummy. While this chapter cannot hope to cover nearly all of the relevant facets of this question, it will be suggested here that a step towards a fuller understanding may be found in the conceptual framework of Maurice Merleau-Ponty's notions of the visible and the invisible and Gilles Deleuze's (e.g. 1994 [1968]) 'ontology of becoming'.

This chapter deals chiefly with the period of ancient Egyptian history known as the Old Kingdom (c. 2700–2200 BCE) with a few examples taken from later times, although many of the more general points could be said to hold throughout the Egyptian history.

The Corpse

The treatment of the corpse in the ancient Egyptian mortuary cult has captured popular imagination well into the present day. This is no doubt due primarily to the fact that the process of mummification entails a transformation blurring one of the most fundamental human distinctions, namely that between the artefact and the person (Meskell 2004: 124–30; Nyord 2009: 1–3). Constituting the inert object of a series of technical procedures of mummification, the dead body is unequivocally devoid of the agency and intentionality characteristic of a human being, and even the appearance of the body is changed into a clearly artificial form. At the same time, however, as a ritual object and a primary locus of personhood of the deceased, the mummy retains much of it humanity, and at least once the procedures are complete, the deceased has become a different type of being with a new kind of agency and embodiment resulting from the ritual reintegration of the deceased in corporeal and social terms (Assmann 2001).

During the mummification process, some central parts of which are constituted by desiccation and removal of body parts particularly prone to decay (Aufderheide 2003: 212–59; Theis 2011: 53–66), the materiality of the corpse changes profoundly from the soft, perishable remains of a living human, to a hard, immobile new shape notionally akin to the cult statue, another ambiguous entity in Egyptian ontology often associated with the mummy in the hieroglyphic writing system (Assmann 2001: 143–5). This change in materiality corresponds to an ontological transformation where what was formerly a living human being goes through a ritual transition to the state of an ancestral spirit, or *akh*.

Thus, rather than regarding the process of mummification as an ensemble of prostheses each making a small improvement to the durability of the dead body, the Egyptian view casts the process as a whole as a transformation of both embodiment and personhood. A ritual text for the reconstitution of the body written in royal pyramids of the Old Kingdom expresses this bodily transformation in terms of materiality:

> O you N (the name of the deceased), raise yourself on your copper bones and your golden members. Your body is that of a god. It cannot moulder, it cannot be destroyed, it cannot putrefy! (Faulkner 1969: 72 §2244a–c)

The divine status of the transfigured deceased is connected here to having an enduring body made of metal not subject to temporal processes of decay. At the same time, the distinction between hard and soft parts of the body expressed here as bones of copper and members of gold reflects a differential treatment in the process of mummification. The soft, perishable parts of the body need particular attention in order to be preserved, while the hard parts may have been regarded as already closer to the immutable divine ideal. Crucially, this difference in materiality ultimately hinges on differences in temporality – the flesh and internal organs are parts which putrefy quickly, while the bones have a much longer cycle of decay. The process of mummification is thus not so much an attempt to preserve the living body as to transform it into a shape materially and temporally suitable for the new form of existence of the deceased.

The timing of the process of mummification itself seems to have been of some importance, as the ideal length of 70 days is mentioned repeatedly in Egyptian texts (Theis 2011: 182–95). This particular period can be connected to the number of days in which the decanal stars that played a crucial role in ancient Egyptian timekeeping and religion were absent from the night sky each year (Theis 2011). One rationale might thus have been for the deceased to be removed, as in the recurrent movement of the stars, from the yearly cycle before passing over to his new mode of existence.

Unlike the case with monuments for the dead in certain other cultures, Egyptian tomb inscriptions only very rarely make reference to the specificities of the manner or time of death of the tomb owner, thus effectively obliterating the particulars of the passage from life to death, an example of the many ways in which the decoration and inscriptions of a tomb were strongly governed by decorum (Baines 1985: 277–305).

Perhaps unexpectedly from this perspective, one does in a number of instances find depictions of the funerary rituals taking place following the death of the deceased (Theis 2011). Such depictions can generally be assumed to have been made well before the actual death and funeral, so that they depict an event as it was imagined or expected to take place at a time subsequent to that of its crafting, while conversely it is understood as a memorandum of an actual past event by the time the tomb is used in the mortuary cult. The combination of the lack of details about the actual death and the emphasis on the ritual transition from human being to ancestor indicate that what was important in the tomb context was the general status of the deceased as an *akh* (effected, as the inscriptions tell us explicitly, by the performance of ritual), rather than the particulars of the life and death of the tomb owner (apart from a few main themes of relevance to the status of an *akh*, as will be seen below).

The Mortuary Cult

The mortuary cult formed the main locus of relations between the living and the ancestors. The central participants of the cult were the close kin of the deceased, although to judge from some of the more detailed representations of cultic services, it could be an extensive affair, including musicians, dancers and offerings of animals as large as oxen. Most of these activities are known from depictions in the tombs themselves (see e.g. Badawy 1981), where they obviously could not serve as commemorations of rituals actually performed, but rather as matrices for the future cult as wished for when the tomb was prepared. As such the depictions contain more than a small amount of idealisation and probably in many cases exaggeration.

The rationale behind the offering cult is rarely spelled out explicitly in Egyptian texts. In the early evidence, the offering ceremony is referred to as 'feeding the *akh*' with a verb otherwise used literally of feeding animals as depicted in Old Kingdom tombs. In this conception, the offerings are thus regarded as food served for the deceased in a manner similar to the feeding of livestock. It has even been suggested (Bolshakov 2009) that the origin of the complex mortuary cult lies in this relatively simple notion of feeding.

In practical terms, the continuation of the cult would have been ensured by the establishment of a funerary estate consisting of fields and personnel known as *ka*-servants (see further below for the concept of *ka*) contractually obliged to perform the cult. In some cases, such contracts have been carved in the tomb itself, providing insight into the legal and practical considerations behind the establishment of the cult (Goedicke 1970).

The Tomb as Setting for Mortuary Ritual

The presentation of offerings took place by the architectural element of the tomb superstructure known in Egyptology as the 'false door', so-called because it 'imitates the most important parts of an Egyptian door, but the niche offers no real entrance to any interior space' (Wiebach-Koepke 2001: 498). Carrying depictions of the deceased and the cult performed for him, the false door forms the focal point of access to the deceased. Notionally, the false door would lead to the inaccessible part of the tomb where statue(s) of the deceased were kept, but its function is also likely to have been predicated originally by the presence of the tomb chamber holding the actual body of the deceased. The exact kind of contact established in the cult is rarely explicated beyond the immediate signification of the door (cf. Meskell 2003: 41, citing Gaston Bachelard's *Poetics of Space*). However, the central ritual recitation called *peret-(er)-ḫeru* refers literally to 'emerging' or 'coming forth at the voice', a notion clearly implying that the deceased comes out from within a bounded space as the offering liturgy is recited, and it is tempting to connect this to the image of the door and the experience of the presence of the dead behind the false door.

Generally, the interplay between presence and absence is based in Old Kingdom tombs on the so-called *serdab* or statue chamber (see Fitzenreiter 2001 for cult statues in the Old Kingdom). The *serdab* houses one or more statues of the deceased, but is completely walled off from the accessible parts of the tomb, leaving just one or two small windows through which visitors would have been able to peek into the dark statue chamber without being able to actually access it nor to get a very clear look at its contents.

Figure 10.1 Section of a typical Old Kingdom *mastaba* tomb showing the *serdab*, false door and subterranean sarcophagus chamber (image by Henrijette Vex Nyord)

In a typical Old Kingdom tomb of the rectangular *mastaba* type, the vertical shaft leading to the tomb chamber in the substructure is accessible from the roof of the tomb, making it completely separate from the rooms accessible horizontally in which the cult was performed. Thus, an experience of the presence of the deceased is established not only through the depictions and inscriptions on the walls of the tombs but also through the architectural elements of the false door, the *serdab* window and the even less concrete awareness of the presence of the body in the separate tomb chamber. These architectonic features assign the presence of the ancestor to particular areas of the tomb, based on the tension between visibility (the door and windows themselves, the half-discerned figure in the *serdab*) and invisibility (the area behind the false door, the *serdab* as a room inaccessible for the living and only partially seen, if at all). These spaces thus become part of the very structure of the tomb, clearly experienced without, however, being seen directly, as expressed by Maurice Merleau-Ponty:

> If it is to reach its perfect density, in other words if there is to be an absolute object, it will have to consist of an infinite number of different perspectives compressed into a strict co-existence, and to be presented as it were to a host of eyes all engaged in one, concerted act of seeing. The house *has* its water pipes,

> its floor, perhaps its cracks which are insidiously spreading in the thickness of its ceilings. We never see them, but it *has them* along with its chimneys and windows which we can see … Taken in itself – and as an object it demands to be taken thus – the object has nothing cryptic about it; it is completely displayed and its parts co-exist while our gaze runs from one to another, its present does not cancel its past, nor will its future cancel its present. (Merleau-Ponty 2002 [1945]: 81, emphasis in original)

The architecture of the tomb, by setting up a door that cannot open and a window through which one almost cannot see, emphasises the interplay between these two ways of experiencing an object, developed by Merleau-Ponty in his later work into an ontological opposition between the visible and the invisible also very much at the heart of pictorial depictions found in the tomb (Nyord fc; cf. also the related use of Merleau-Ponty's notion of a 'view from everywhere' as basis for perspectivism in Willerslev 2011).

Another way to view this 'localised absence' is to understand the existence of the deceased in the tomb in terms of *intensities* as understood by Gilles Deleuze:

> Intensity is simultaneously the imperceptible and that which can only be sensed. How could it be sensed for itself, independently of the qualities which cover it and the extensity in which it is distributed? But how could it be other than 'sensed', since it is what gives to be sensed, and defines the proper limits of sensibility? (1994 [1968]: 290)

In this perspective, the person while he was alive would have been conditioned by syntheses of such intensities, but upon the transition to his new form of existence in the tomb, the ancestor reverts to a condition of pure intensities. Deleuze (1994 [1968]: 322) describes this aspect of death as an 'internal power which frees the individuating elements from the form of the I or the matter of the self in which they are imprisoned'. The Egyptian designation of such a virtual being (akin to the Amazonian spirits discussed by Viveiros de Castro [2007: 157–8]), is the *ka* of a person (see also Assmann 2011: 137–43 for the *ka* in relation to time). Correspondingly, upon death a human being is said to 'go to his *ka*', which can be understood as a return to a fully virtual existence. Similarly, the offerings presented to the deceased in the tomb are explicitly (and regularly so from the Middle Kingdom on, cf. Fitzenreiter 2001: 547) destined for his *ka* as the predominant virtual form of existence of the ancestor. Precisely as one would expect from this interpretation of the *ka* as the virtual side of which the actual person is a manifestation, it also plays a role conditioning the becoming and behaviour of the living human being (Junge 2003: 122–6), corresponding to Deleuze's notion of the real individual as driven by configurations of virtual intensities. In Egyptian, this virtual conditioning of the actual is expressed in personified metaphors e.g. of the *ka* 'abhorring' particular behaviour that should be avoided (see Junge 2003: 123 for examples).

This almost structural presence of the ancestor within the tomb building is further emphasised in its inscriptions. Thus, addresses to the tomb visitors frequently make reference to acts that can broadly be characterised as threatening the integrity of the tomb as dwelling place of the ancestor, referred to generally as 'doing bad things against this (building)' (Morschauser 1991; for this and the following phrases, see Edel 1944: 3–30). This can be thought of in the very concrete sense of 'removing any stone from this tomb of mine', but also in terms of various ritual transgressions, summarised as 'entering this tomb of mine in a state of impurity', particularly 'after having eaten taboos which are abhorred by an ancestral spirit (*akh*)', but also occasionally other sources of impurity. Such states of impurity should be cleansed by ritual purification before entering the tomb, as inscriptions refer to the way visitors 'should purify themselves for an able ancestral spirit'. These descriptions are followed by threats that the deceased will effectuate against transgressors, but the inscriptions do not dwell on the potential ill consequences for the deceased of such violations beyond the clear implication that they would compromise the tomb as a place of dwelling and manifestation of the ancestor.

Designations of the wider context of the tomb generally corroborate this view of a dwelling-place of the ancestor. Thus, the necropolis is commonly designated *ẖery-netjer*, 'the domain of the god', where the 'god' probably initially refers to the king, around whose pyramid the tombs of his courtiers were originally clustered (Allen 2006: 11). More rarely, the necropolis (and hence, by extension, the realm of the dead) is referred to as a 'city' (*niwt*), as when a tomb owner refers to building his tomb 'in the city of my lord', once again incorporating the emphasis of the royal tomb as centre of the necropolis (Allen 2006: 10 n. 7), although other social principles were also at work in the layout of necropoleis in clusters of tombs (Baines and Lacovara 2002: 9–10).

Egyptian Conceptions of Time

The conceptual framework current in Egyptology understands Egyptian conceptions of time as revolving around the two central notions of *neḥeḥ* and *djet* (recent discussions by Servajean 2007; Meyer-Dietrich 2009; Assmann 2011). The two concepts are understood as complementary with *neḥeḥ* dealing with time as cyclical, although the two most recent monographic treatments of the question disagree as to whether, on this basis, it is best translated as 'eternity' (Assmann 2011) or simply as 'time' (Servajean 2007). The notion of *djet*, on the other hand, expresses permanence and immutability. Correspondingly, *neḥeḥ* is particularly connected to the flux of human life, although it is also characteristic of the various cosmic cycles playing an important role in ancient Egyptian agriculture, timekeeping and religion, such as the cycles of the celestial bodies, the yearly flooding of the Nile, etc. In contrast, *djet* is understood as particularly connected with the sphere of existence of the gods and the dead. The places where these beings are primarily encountered, the temple and the tomb, respectively, are characterised precisely by the *djet*-aspect inherent in their changeless monumental form and lithic materiality. With reference to the dominant examples of each of the two modes of existence, Assmann (2011) coins the terms 'Steinzeit und Sternzeit' to illustrate the Egyptian concepts.

In this perspective, the material transformation from dead body to quasi-personal artefact (mummy) thus reflects not only the transformation from human to ancestor, but also the passage from existence in one sphere of time to another. While living, humans are fundamentally subjected to temporal cycles of decay and regeneration: the daily, monthly and annual cycles of nature, and the birth-death cycle of passing generations, subsumed under the Egyptian term *neḥeḥ*. Upon death, the human being partakes in the new, complementary form of static, unchanging existence, of which the incorruptible mummy and the stone-built tomb by virtue of their material attributes offer prototypical examples, designated by the term *djet*.

It should be noted at this point that the most important previous studies of the two terms have been based methodologically on a top-down approach where individual occurrences of the two words have been used as illustrations of a hypothesis, rather than categorising and analysing the patterns of such occurrences. This effectively situates the current Egyptological understanding as detailed above somewhat uneasily between the status of *emic* and *etic* terminology. This is not the place to attempt to remedy this situation, and as can be seen even from the brief remarks made here, the traditional understanding of the concepts offers a great deal of explanatory power as analytical terms, but whether they are exactly what the Egyptians understood (at a given moment or throughout the history of the language) by the words *neḥeḥ* and *djet* must remain an open question. For this reason, it is useful to supplement the notions of *neḥeḥ* and *djet* with additional analytical perspectives as will be exemplified in the next section.

In any case, the importance of the two concepts for the Egyptian understanding of the function of funerary and commemorative monuments can be illustrated by a passage from an inscription dealing with the restoration of an ancient cult place dedicated to the deified noble Heqaib. The person responsible for the restoration describes the deplorable state in which he found the monument in the following words:

> The earth had swallowed its sanctuary. *Neḥeḥ* was not foreseen, as it was like the work of a foreigner who does not recall things with a view to *djet* (or: towards *djet*, i.e. 'ever'). (Habachi 1985: 36, F. 3, pl. 24, l. 8–9)

This passage shows that both *neḥeḥ* and *djet* are conceptualised as directions of orientation of the cultic monument, and also that the author of the inscription regards this notion of monumentality and cult with an inherent view to eternity as a particularly Egyptian trait. It also serves to illustrate the kind of parallelism often found between the two concepts which further complicates the task of making a precise distinction between them.

Syntheses of Time in the Mortuary Cult

The tomb and the mortuary cult taking place there incorporate many less explicit references to time. The tomb itself is an obvious result of the past in more than one sense. Most concretely, the tomb, as any permanent building, is the end result of work carried out in the past. This is doubly true of the Egyptian tomb, however, because the presentation of the tomb owner as someone who lived in the past plays a central role in the texts and images of the tomb. They present this life in a fossilised form intended precisely to represent him at a point in time when he is no longer the living human being presented in the autobiography, but has shifted his ontological status, as other parts of the texts stress repeatedly, to that of an ancestral spirit.

This presentation of the past in the inscriptions has two important aspects, corresponding to two different parts of the traditional tomb inscription, known in Egyptology as the *narrative autobiography* and the *ideal (or ethical) autobiography*, respectively (cf. Kloth 2002: 221–57). The first type of inscription presents an overview of the life of the tomb owner in the service of the king, emphasising his successes and the corresponding favours and boons granted him by the king. This part of the tomb's inscriptions can be remarkably detailed with dates, names and places, especially when describing the major projects forming the highlights of the tomb owner's career, such as important expeditions. An example of such an inscription may be found in the celebrated and highly detailed description of the expeditions of the nobleman Harkhuf in his tomb near Aswan in southern Egypt from the middle of the 6th Dynasty (c. 2300 BCE). This part of the inscription begins with a characterisation of the first expedition of Harkhuf, undertaken under the leadership of Harkhuf's father, but characteristically still presented from the perspective of the tomb owner himself:

> The Incarnation of king Merenre (2285–2279 BCE), my lord, sent me along with my father, the Sole Companion and lector priest, Iri, to Iam to open the route to that foreign land … His Incarnation sent me a second time, this time alone. I went up by the Elephant road and returned by way of Irtjet, Makher, Teres and Irtjetj in the span of eight months. When I came back, I was bringing a great amount of goods from those foreign countries, such as had never been brought to this land before. (Sethe 1933: 124,13–125,7; translation Strudwick 2005: 330)

It will be noted that Harkhuf presents himself characteristically as superseding his predecessors, his deeds being unsurpassed at his time. In this way, the details of an individual's life become moulded in the ideal curriculum vitae of a loyal liegeman of the king. Another characteristic feature is that only very general knowledge (such as familiarity with kings and basic geography) is taken for granted in the inscription, and new persons, such as Harkhuf's father, are introduced with names and titles. This entails in principle that no personal acquaintance with the described events is presupposed in the reader, as one might have expected if the main readership is the family and associates of the tomb owner. Thus, the presentation does not implicate the memory of the reader, offering an account of the tomb owner's life that can be read divorced from any particular time or social setting (in partial contrast to the case with the domestic shrines studied by Meskell [2003]).

While all parts of the decoration of the tomb are governed by decorum, the other type of autobiography, explicitly labelled the *ideal autobiography*, consists mainly of a selection from a set of stock phrases describing the behaviour of a good official. This part of the inscription contains none of the particulars of the life of the tomb owner, deliberately casting him as little more than an exemplification of a prototype of virtuous behaviour.

> Today I have gone up from my town and descended from my district, having built my house and set up doors, having dug my pool and planted trees. My king favoured me, and my father made a will for me. I am an excellent … beloved by my father, praised by my mother, one loved by all my siblings. I gave bread to the hungry and clothes to the naked and brought to land the one who had no ferryboat. (Sethe 1933: 121,11–122,8; translation Strudwick 2005: 329)

Although the first part of this passage may not seem all that different from the description of Harkhuf's expeditions above, these sentences are in fact entirely formulaic, repeated with minor variation in a great number of tombs (Kloth 2002: 54–128), and serving to capture the invariant elements of the preparation of a tomb complex before finally leaving the land of the living in favour of the new form of existence in the tomb. The second half of the quoted passage is more clearly formulaic, giving statements of what is essentially not just particularly Egyptian, but cross-cultural, ideals of ethical behaviour in a nobleman.

While both of these ways of presenting the tomb owner are obviously idealised, they also serve two different purposes. The Egyptologist Jan Assmann (1991: 186–7) has formulated this difference as the first type of autobiography dealing with 'historical discourse', while the second deals with 'wisdom' (a designation to be understood in the light of the particular Egyptian genre of 'wisdom literature' presenting ideals of ethical and otherwise correct behaviour).

From the point of view of questions of time, however, both ways of presenting the deceased as a being of the past obviously refer to the same point in time (namely the duration of the tomb owner's life), although the second type of text takes a more holistic and less sequential view than the first. One way to conceptualise these two different views of the past is by reference to Gilles Deleuze's fundamental discussion of syntheses of time in chapter II of *Difference and Repetition* (1994 [1968]: 91–156; cf. Turetzky 1998: 211–29; Hodges 2008; Williams 2011).

The first mode of presentation is essentially an enumeration of a sequence of passing presents, some of which can be of significant duration in the narration (e.g. an eight-month journey through a succession of localities). Many significant events in the life of a human are omitted in this sequence, as they are of no significance to the image of a professional career of servitude for the king dictated by decorum as the contents of autobiographical tomb inscriptions. Such a succession of presents is the kernel of what Deleuze labels the first synthesis of time, a passive synthesis where the past and future are contracted as dimensions of the present (Deleuze 1994 [1968]: 91). At the same time, however, there is a sense of unity that makes all of these instants belong to 'the same life', a unity that Deleuze refers to as 'destiny', and in a sense this is what it is attempted to capture in the ideal biography. Of this notion, Deleuze (1994 [1968]: 105) writes:

> Destiny never consists in step-by-step deterministic relations between presents which succeed one another according to the order of a represented time. Rather, it implies between successive presents non-localisable connections, actions at a distance, systems of replay, resonance and echoes, objective chances, signs, signals and roles which transcend spatial locations and temporal successions.

These commonalities, connections or underlying tendencies are precisely what it is attempted to capture in the formulaic expressions of the kind 'giving bread to the hungry and clothes to the naked'. Unlike the narrative autobiography, these ideal expressions do not refer to particular incidents, but rather to what unites or underlies the events of the tomb owner's life:

> The succession of presents is only the manifestation of something more profound – namely the manner in which each continues the whole life, but at a different level or degree to the preceding, since all levels and degrees coexist and present themselves for our choice on the basis of a past that was never present. (Deleuze 1994 [1968]: 105)

This 'more profound' owes its existence to the 'pure past' in Deleuze's second synthesis of time. Unlike the past as dimension of the present in the first synthesis of time, the pure (or virtual, or *a priori*) past has never been present, but instead acts as a ground and is what causes the present, as the most contracted level of the pure past, to pass.

Thus, the difference between the two types of autobiography is that one presents the more superficial particular details of the professional career of the tomb owner with its causal and sequential relations, while the other attempts to characterise the more profound, underlying relations of which those details are the actualisation. This has a couple of important entailments for the role of each part of the inscription.

First, the fact that the ideal autobiography is shared between tomb owners to a significant degree due to its formulaic nature, serves to present such tomb owners as essentially alike, despite the differences revealed in other parts of the tombs' inscriptions such as the titles and responsibilities of individual deceased, not to mention the events and deeds shaping each individual's life. While such differences might at first sight seem much more prominent, the perspective of different syntheses of time taken here, where the presentation based on the first synthesis gives a more contingent, more superficial image than that based on the second, helps to explain why the phraseology of the ideal autobiography proved remarkably resistant to change during the history of ancient Egypt.

The second entailment is that whereas the precise role of the sequential autobiography in the greater whole of the tomb complex and the mortuary cult is more difficult to pinpoint (cf. the discussion in von Lieven 2010), the ideal autobiography often stands in a direct connection with the kind of appeals to tomb visitors referred to above admonishing them to behave appropriately in the tomb. This in effect makes the ideal biography function as a manner of argument or justification why the visitor should heed the warnings and requests of the tomb owner. Thus while the more contingent facts of the tomb owner's life rarely seem to play any distinct role after his death (except as an illustration of the qualities he claims to have), the underlying connections and tendencies apparently remain of great importance in the relations between the tomb owner in his new mode of being and the living.

Deleuze's two syntheses of time as presented here can also be employed to deepen our understanding of the entailments of the Egyptological reading of the notions of *djet* and *neḥeḥ*. Thus, the general tendency in Egyptological treatments to understand *djet* as referring especially or exclusively to affairs of the 'beyond' (e.g. Assmann 2011: 84–5) would accord well with a connection to the pure past in Deleuze's second synthesis of time, while the kind of repetition connoted by *neḥeḥ* is reminiscent of the succession of presents in the first synthesis (and both notions would be involved in the return of pure difference as the condition of the new in the third synthesis discussed below). As was pointed out above, however, much work remains to be done on our understanding of the two Egyptian terms, and for the moment, these brief remarks can be no more than an indication for a possible direction of future research into their complexities.

In addition to these perspectives on the past life, the future existence of a deceased individual is dependent, as described above, on the transformation into the permanent shape of a mummy. This notion of an unchanging materiality seems to run counter to all experience of the entropy of things. In the words of Henri Bergson:

> The object may remain the same, I may look at it from the same side, at the same angle, in the same light; nevertheless the vision I now have of it differs from that which I have just had, even if only because the one is an instant older than the other. My memory is there, which conveys something of the past into the present. (Bergson 2007 [1907]: 1)

The notion of looking at an object from every side is reminiscent of Merleau-Ponty's notion of the invisible as referred to briefly above in connection with the hidden chambers of the tomb, but with the addition that experience never offers an unchanging object. The ideal of the unchanging mummy and tomb belies this basic experience to establish an eternal unchanging mode of being. In practice, of course, experience will occasionally betray this ideal, and certain Egyptian texts point for example to the perishable nature even of tomb buildings of stone – thereby calling into question the whole notion of the ancestor's continued existence in the tomb predicated on it (Baines and Lacovara 2002: 17).

Apart from such occasional sceptic voices, several different types of sources attest to the notion that the ancestors would use the special abilities entailed by their status as an *akh* to help and support those who fulfilled the ancestor's wishes and needs in the mortuary cult. The tomb inscriptions illustrating this notion not unexpectedly concern wishes of a very general nature. An example of such promises from the Middle Kingdom (where they were often addressed to general passers-by rather than specific people socially or contractually obliged to carry out the cult) reads as follows:

> The king is rejuvenated in your lives, the monuments of your local gods shall endure for you, you shall be in the favour of your sovereign, you shall pass on your offices to your children, your offspring shall be firm in your seats in your offices permanently (*djet*), you shall not go hungry, you shall not go thirsty. The great god has commanded that you exist on earth in his favour. You shall not be incarcerated in an irksome place, due to the favour of your local gods, when only you say 'A royal offering for Osiris, Lord of Abydos, the Great God Wenen-nefer: A thousand bread and beer, beef and fowl, an invocation offering at every festival for the *ka* of the keeper of the diadem, liegeman of the Great House, Nebipusenusret, begotten by Ita.' For the breath of the mouth is beneficial for the noble dead, it is nothing wearisome, but you shall be an Imperishable One, a star within She-of-a-Thousand-Akhu. (Budge 1912: pl. 2, translation Parkinson 1991: 139–42)

As will be seen, many such wishes are of a very general nature, and with their stress on royal and divine favour, professional accomplishment, and even success beyond death (as a star in the sky), they form a kind of mirror image in the future and the second person of the idealized life described in autobiographies.

More concrete wishes for the future would naturally have to be formulated by the other side in the interaction, namely by the living. Such wishes were occasionally formulated in writing in the so-called Letters to the Dead deposited in tombs, which provide insights into the relations between the dead and the living (see Wente 1990: 211–19).

For the present purposes one particular aspect of such interactions will serve adequately as an example of the potentiality inherent in the ancestral existence in the tomb, namely the notion that the deceased causes (directly or indirectly) children to be born in his surviving family (cf. Nyord 2009: 424–59). Central in Deleuze's third synthesis of time is the eternal return, and particularly the notion that what returns is not the same, but rather pure differences. The third synthesis of time consists of a reorganisation of time around a *caesura* where a metamorphosis takes place, and which cuts up time asymmetrically into a before and an after. Deleuze (1994 [1968]: 117) expresses the relation of this synthesis of the future to the first two syntheses of time as follows:

> The future, which subordinates the other two [i.e the past and the present in the first two syntheses] to itself and strips them of their autonomy, is the royal repetition. The first synthesis concerns only the content and the foundation of time; the second, its ground; but beyond these, the third ensures the order, the totality of the series and the final end of time.

In this way, the third synthesis is the condition for the new complementing the first synthesis of habit and the second synthesis of memory. What is at work in the various Egyptian cases of 'help' from beyond the grave can be seen in this perspective as the return of intensities of the kind suggested above to be inherent in the figure of the *ka*.

A figurine of Middle Kingdom date of a woman carrying a child on the hip (Schott 1930), while belonging to a well-attested type of figurine deposited primarily at tomb and temple sites (Pinch 1993), is highly unusual in carrying an inscription specifying its purpose, namely 'May a child-birth be given to your daughter Seh'. The figurine was thus meant to occasion the situation that it depicts (a woman with a child), and its deposition in the tomb of the father of the woman depicted was considered integral to this effect.

A number of interesting consequences can be drawn from this artefact (Nyord 2009: 448–53; cf. also Meskell 2004: 69–73 for votives dedicated to ancestors more generally), but for the present purposes, it is the notion of the deceased causing or being otherwise involved in the conception and birth of a child that is of particular relevance. In Egyptian texts the lines are often blurred about the precise role of the deceased and/or his *ka* in this connection (cf. Nyord 2009: 448–53). One unusually clear expression in a wisdom text employs the metaphor of the *ka* (in this case of a living person) 'begetting' a child who takes after his father (Žába 1956: 32, l. 204) – with the interesting corollary that this is emphatically *not* the case with a son who is incompetent and unsuccessful, thereby showing no intrinsic relation to his father.

The function of the figurine can then be said to be to serve as a matrix for individuation in Deleuze's (1994 [1968]: 308) sense of 'the act by which intensity determines differential relations to become actualized'. To accomplish such a change (in this case conception of a child) in a living person, it is necessary to access the virtuality of pure difference, available in the Egyptian world primarily in the tomb and the temple, and it is likely, at least in earlier times, that the specific locus of the ancestral tomb played a role of particular importance (Nyord 2009: 453). This is accomplished by the establishment of a caesura at the ritual deposition of the figurine whereby it is sought to bring about the desired metamorphosis, ordering time into a before and after the decisive event of the conception of a child. It is interesting to note in this regard that another example of a petition to an ancestor for childbirth (in this case from a letter to the dead, not a figurine) further specifies the intensities involved by wishing for 'a healthy male child' (Gardiner 1930: l. 4 and 8a).

Conclusion

While the particularities of the individual death are rarely admitted into the tomb's programme of decoration and inscription, the mummified and hidden dead serves as a crucial nexus for the gathering of past, present and future in the ancient Egyptian tomb. As a mode of being characterised by the notion *djet*, the mummy is immobile and changeless, but at the same time a locus for creative potentials that can be actualised. Thus, the mummy, as the product of a life lived in the past, now in the present fixed shape, offers promise of the actualisation of new creations. The mortuary cult serves as the activity that brings the living into contact with this illusive being, where the times of the mummy are reflected as remembrance of a past life, present instantiations of the offering cult, and the wishes for (or threats of) the renewed actualisation of the potentialities hidden in the tomb chamber. In this way, the ancient Egyptian mortuary cult offers a striking example of death and the dead as a central factor in human conceptions of time.

References

Allen, J.P. 2006. Some aspects of the non-royal afterlife in the Old Kingdom, in *The Old Kingdom Art and Archaeology. Proceedings of the Conference Held in Prague, May 31–June 4, 2004*, edited by M. Bárta. Prague: Publishing House of the Academy of Sciences of the Czech Republic, 9–17.

Assmann, J. 1991. *Stein und Zeit: Mensch und Gesellschaft in alten Ägypten*. Munich: Wilhelm Fink.

Assmann, J. 2001. *Tod und Jenseits im alten Ägypten*. Munich: C.H. Beck.

Assmann, J. 2011. *Steinzeit und Sternzeit: Altägyptische Zeitkonzepte*. Munich: Wilhelm Fink.

Aufderheide, A.C. 2003. *The Scientific Study of Mummies*. Cambridge: Cambridge University Press.

Badawy, A. 1981. The spiritualization of Kagemni. *Zeitschrift für Ägyptische Sprache und Altertumskunde* 108, 85–93.

Baines, J. 1985. *Fecundity Figures: Egyptian Personification and the Iconography of a Genre*. Warminster: Aris & Philips.

Baines, J. and Lacovara, P. 2002. Burial and the dead in ancient Egyptian society: respect, formalism, neglect. *Journal of Social Archaeology* 2(1), 5–36.

Bergson, H. 2007 [1907]. *Creative Evolution*. Trans. Arthur Mitchell. Basingstoke and New York: Palgrave Macmillan.

Bolshakov, A. 2009. Три этюда по организации культа при IV династии [Three notes on the organization of the cult in the Fourth Dynasty], in *St. Petersburg Egyptological Readings 2007–2008 in Commemoration of Oleg Dmitrievich Berlev On the Occasion of his 75th Birthday: Papers of the Conference*. Transactions of the State Hermitage 45. St. Petersburg, 25–39.

Budge, E.A.W. 1912. *Hieroglyphic Texts from Egyptian Stelae &c., in the British Museum: Part II*. London: British Museum.

Deleuze, G. 1994 [1968]. *Difference and Repetition*. Trans. Paul Patton. London and New York: Continuum.

Edel, E. 1944. Untersuchungen zur Phraseologie der ägyptischen Inschriften des Alten Reiches. *Mitteilungen des Deutschen Instituts für Ägyptische Altertumskunde in Kairo* 13, 1–90.

Faulkner, R.O. 1969. *The Ancient Egyptian Pyramid Texts: Supplement of Hieroglyphic Texts*. Oxford: Oxford University Press.

Fitzenreiter, M. 2001. *Statue und Kult: Eine Studie der funerären Praxis an nichtköniglichen Grabanlagen der Residenz im Alten Reich*. Internet-Beiträge zur Ägyptologie und Sudanarchäologie 3. Berlin: Humboldt-Universität zu Berlin. Available at: http://www2.rz.hu-berlin.de/nilus/net-publications/ibaes3 [accessed: 12 January 2012].

Gardiner, A.H. 1930. A new letter to the dead. *Journal of Egyptian Archaeology* 16, 19–22.

Goedicke, H. 1970. *Die privaten Rechtsinschriften aus dem Alten Reich*. Beihefte zur Wiener Zeitschrift für die Kunde des Morgenlandes 5. Vienna: Notring.

Habachi, L. 1985. *Elephantine IV: The Sanctuary of Heqaib*. Archäologische Veröffentlichungen 33. Mainz am Rhein: Philipp von Zabern.

Hodges, M. 2008. Rethinking time's arrow: Bergson, Deleuze and the anthropology of time. *Anthropological Theory* 8(4), 399–429.

Junge, F. 2003. *Die Lehre Ptahhoteps und die Tugenden der ägyptischen Welt*. Orbis Biblicus et Orientalis 193. Freiburg and Göttingen: Universitätsverlag Freiburg Schweiz, Vandenhoeck & Ruprecht.

Kloth, N. 2002. *Die (auto-)biographischen Inschriften des ägyptischen Alten Reiches: Untersuchungen zu Phraseologie und Entwicklung*. Beihefte Studien zur Altägytischen Kultur 8. Hamburg: Buske Verlag.

von Lieven, A. 2010. Zur Funktion der ägyptischen Autobigraphie. *Die Welt des Orients* 40, 54–69.

Merleau-Ponty, M. 2002 [1945]. *Phenomenology of Perception*. Trans. C. Smith. London and New York: Routledge.

Meskell, L. 2003. Memory's materiality: ancestral presence, commemorative practice and disjunctive locales, in *Archaeologies of Memory*, edited by R.M. Van Dyke and S.E. Alcock. Oxford: Blackwell, 34–55.

Meskell, L. 2004. *Object Worlds in Ancient Egypt: Material Biographies Past and Present*. Oxford and New York: Berg.

Meyer-Dietrich, E. 2009. Herr der Ewigkeit, der die Jahre der Götter aufgehen lässt, und der, in den Re täglich hinabsteigt: zur Frage altägyptischer Theologie und Zeitvorstellungen, in *Le jour de Dieu = Der Tag Gottes : 5. Symposium Strasbourg, Tübingen, Uppsala, 11. – 13. September 2006 in Uppsala*, edited by A. Hultgård and S. Norin. Wissenschaftliche Untersuchungen zum Neuen Testament 245. Tübingen: Mohr Siebeck, 1–18.

Morschauser, S. 1991. *Threat-Formulae in Ancient Egypt: A Study of the History, Structure and Use of Threats and Curses in Ancient Egypt*. Baltimore: Halgo.

Nyord, R. 2009. *Breathing Flesh: Conceptions of the Body in the Ancient Egyptian Coffin Texts*. Copenhagen: Museum Tusculanum.

Nyord, R. Forthcoming. Vision and conceptualization in ancient Egyptian art, in *Sensuous Cognition: Explorations into Human Sentience: Imagination, (E) motion and Perception*, edited by R. Caballero and J.E. Díaz-Vera. Berlin: Mouton, 136-168.

Parkinson, R.B. 1991. *Voices from Ancient Egypt: An Anthology of Middle Kingdom Writings*. London: British Museum.

Pinch, G. 1993. *Votive Offerings to Hathor*. Oxford: Griffith Institute.

Schott, S. 1930. Die Bitte um ein Kind auf einer Grabfigur des frühen Mittleren Reiches. *Journal of Egyptian Archaeology* 16, 23.

Servajean, F. 2007. Djet *et* Neheh*: Une histoire du temps égyptien*. Orientalia Monspeliensia 18. Montpellier: Université Paul Valéry-Montpellier III.

Sethe, K. 1933. *Urkunden des Alten Reiches*. Leipzig: J.C. Hinrichs.

Strudwick, N. 2005. *Texts from the Pyramid Age*. Writings from the Ancient World 16. Atlanta: Society of Biblical Literature.

Theis, C. 2011. *Deine Seele zum Himmel, dein Leichnam zur Erde: Zur idealtypischen Rekonstruktion eines altägyptischen Bestattungsritual*. Studien zur Altägyptischen Kultur Beiheft 12. Hamburg: Buske Verlag.

Turetzky, P. 1998. *Time*. London and New York: Routledge.

Viveiros de Castro, E. 2007. The crystal forest: notes on the ontology of Amazonian spirits. *Inner Asia* 9, 153–72.

Wente, E. 1990. *Letters from Ancient Egypt*. Writings from the Ancient World 1. Atlanta: Society of Biblical Literature.

Wiebach-Koepke, S. 2001. False door, in *The Oxford Encyclopedia of Ancient Egypt*, edited by D.B. Redford. Oxford: Oxford University Press, 498–501.

Willerslev, R. 2011. Frazer strikes back from the armchair: a new search for the animist soul. *Journal of the Royal Anthropological Institute (N.S.)* 17, 504–26.

Williams, J. 2011. *Gilles Deleuze's Philosophy of Time: A Critical Introduction and Guide*. Edinburgh: Edinburgh University Press.

Žába, Z. 1956. *Les Maximes de Ptaḥḥotep*. Prague: Académie Tchécoslovaque des Sciences.

Chapter 11
Death and *Meat*eriality

Kristian Bjørkdahl and Karen Lykke Syse

In his famous tome, *Physiologie du Goût, ou Méditations de Gastronomie Transcendante* (1848), the French lawyer, politician and *gastronome* Jean Anthelme Brillat-Savarin wrote, "Dis-moi ce que tu manges, je te dirai ce que tu es" (ix). With time and movement across borders this saying—"Tell me what you eat and I will tell you what you are"—was condensed and modified into our contemporary proverb, "You are what you eat."

The famous Frenchman already knew what contemporary sociologists of food keep reminding us, namely that food functions as a marker of identity and, thus, of social status (see Scholliers 2001; Counihan and Van Esterik 2008). He already knew that our ways with food reveal who we are, how we live and what we care about.[1] Our food practices form a nexus where technology and economy meet culture and social relations, and hence, to understand how a particular historical epoch eats is to understand much more besides food. Because so many aspects of our lives are written into what we eat, food can be a particularly rewarding object of study for those who want to understand change.

At the same time, many strands of continuity run through the history of food practices. Perhaps the most basic is that food often involves the taking of a life and the treatment of what one would otherwise call a corpse. To eat meat, we must kill. Someone has to die so that we can live. The fact that we take the life of an animal to sustain our own, presents us with a quandary: the killing must be put into a context that makes animal deaths acceptable and meaningful. The living, breathing creature must be transformed—not just physically, but symbolically—into inert matter fit for consumption. The animal must be made edible (see Vialles 1988).

We can thus start from the supposition that every culture relies on a distinctive strategy of symbolic transformation that makes killing and consuming animals acceptable and meaningful. We have examples of such transformation from various historical contexts. We know it from the medieval Catholic Church, where the goose—which lived in water and was believed to hatch through barnacles

1 This is to say much the same as what Roland Barthes said, when he spoke of food as a system of communication. "No doubt, food is, anthropologically speaking (though very much in the abstract), the first need," wrote Barthes, "but ever since man has ceased living off wild berries, this need has been highly structured. Substances, techniques of preparation, habits, all become part of a system of differences in signification; and as soon as this happens, we have communication by way of food" (Barthes 2008: 29–30).

on driftwood—was conveniently defined as a fish during Lent and other days of abstinence (Thomas 1984: 88). We know it from Stephen Hugh-Jones' ethnography of the fish trap used among the Barasana, where "One moment the fish is placidly swimming along belonging (so it thinks) to the animal kingdom and then, bang before it knows what has happened it is a vegetable, dangling from the branches of a tree to be plucked like any other fruit by a passing Indian" (Gell 1999: 231). And we know it from some present-day Western societies, where indistinct parts of a chicken carcass are transformed into an oval object called "nugget."

In this chapter, we turn toward our contemporary ways of dealing with this transformation. If it is true that we are what we eat, and further, that *we are how we kill what we eat*, then what are we, the twenty-first century citizens and consumers of wealthy liberal democracies? We argue that the animal that becomes our meat has been disappearing from the view of the consumer, and that modern Western societies' strategy for solving the old dilemma of killing and eating living creatures now is to suffice without a strategy. In other words, we have no effective way of contextualizing and justifying the transformation of animal to meat—none, that is, except avoidance or concealment.

We argue that the latter part of the twentieth century was marked by an increasing incapacity of Western publics to acknowledge that meat always comes from a living animal which has been killed *to be eaten*. In place of such acknowledgment we have entered mechanisms of concealment that allow us to continue eating dead animals, but only because those same mechanisms obfuscate what we are eating. We no longer eat meat, the animal. Instead we eat meat, the ingredient.

This urge to conceal appears motivated by what we call *biocentrism*, that is, the growing tendency to grant animals a moral standing similar to that of humans. This process explains not just why we have such trouble acknowledging animal deaths; it might also help explain what some have referred to "denial of (human) death" (Becker 1973). There are two sides to biocentrism: on the one hand, it means upgrading of animals, on the other, it means downgrading of humans. To upgrade animals is to put meat eating into a morally troubling zone, where there appears to be only a very thin line between carnivorism and cannibalism. But to downgrade humans is to challenge our traditional view that humans are something more than matter; to die means now simply to rot in the ground.

The Culture of Cookbooks

How can we shed light on the changes that have taken place in our ways of dealing with animal deaths? Most Western people, whether carnivorous or not, encounter meat practically every day. We see it in shops or in our fridge, we smell it cooking in restaurants or at home, we hear it sizzle in frying pans and simmer in pots. Animal flesh is everywhere, and by pivoting an analysis around an everyday foodstuff we can reach a deeper insight into everyday life in the post-war period.

In the following, we apply a cultural historical perspective on this recent past, and do a small empirical study of Norwegian cookbooks to substantiate the claim that the animal and animal-to-meat process is disappearing from the view of the consumer.[2] We argue that, in this period, the animal that becomes our meat largely disappears from cookbooks, and the consumer is left with meat as just another ingredient—not much different from vegetables or pasta. If meat is still considered edible, the animal source of that meat has become distasteful.

In most Western countries, the publication of cookbooks has multiplied in the post-war era. There are now books that deal with different geographical regions, different diets and different foodstuffs. This new plethora of cookbooks can be seen as indicative of any number of societal changes in the twentieth century: various technological advances, the movement of women into the workforce and the subsequent decline of food traditions; the globalization of food production, food distribution and food cultures; and the changing use of food as a social marker.

In this chapter, we want to highlight another change that can be traced in cookbooks, and which we believe is telling of the post-war era, namely the changing visual and textual representation of animals and animal deaths. Our approach is based on the assumption that cookbooks are containers not just of recipes, but of normativity, in the sense that they tell us *what* to eat, *how* to prepare it, as well as *what context* to place it in. As artifacts of everyday life, cookbooks can be understood as repositories of past practices, and are hence prime materials with which to study changing conceptions of meat.

In *Invention of the Modern Cookbook*, historian Sandra Sherman underlines this point when she argues that the cookbook, from its inception as a modern genre in the eighteenth century, was "a complex enterprise that turned their eighteenth-century authors into pitchmen, psychologists, cultural arbiters, and cultural authorities" (Sherman 2010: xi). She points out that, "Modern cookbooks adopt ideological or at least pedagogical positions … [and] opine on everything from saving the planet to controlling one's time":

> they advocate for fresh/local/seasonal cuisine, for veganism, for a return to an alimentary regime consistent with sustainability. They suggest that "thirty-minute meals" equal "cooking," and that childhood indulgences (cupcakes) can be redesigned as elaborate grown-up fantasy. Even the simplest cookbooks have subtexts, advancing ideas about how to shop (farmers markets, not supermarkets) and how many hors d'oeuvres should replace dinner. Cookbooks are complex, multivalent texts, situating food at the center of larger concerns. (Sherman 2010: 229)

2 It is an open question how far our findings extend beyond the Norwegian case. The Norwegian editions of these books have Swedish originals and Danish equivalents, so it seems safe to assume that something like our argument applies in the whole of Scandinavia. We believe that it applies to an even larger area, however, and estimate that the Norwegian case is probably not unlike the situation in Western Europe in general, in Great Britain and North America.

In this chapter, the larger concern at stake is how we deal with animal deaths in the contemporary West. In order to explore this topic and provide an example of the process we believe has taken place in recent decades, we have selected as our empirical example two editions of a comprehensive Norwegian cookbook, *Gyldendals Store Kokebok* (*Gyldendal's Comprehensive Cookbook*). Since it first appeared, this book has been republished in many different reprints and editions. From the first edition in 1955 to the latest edition published in 2002, some chapters have changed radically, while others have received no more than a facelift, most importantly through more international recipes and updated graphic layout. However, what all editions have in common (which to some extent distinguishes this book from many others on the market) is its ambition to function as a comprehensive publication—the Bible of the kitchen. And just as the Bible, *Gyldendals* did not appeal to a particular niche, but to the great majority of the population. The point of the book was to provide answers to any question one could possibly have about how to prepare appetizing and nutritious meals for an entire family. As such, the different editions of this book arguably represent an established and culinary authoritative view of animals and meat, and of the necessity of animal death that stands between those two. Let us begin by opening the meat chapter of the first edition.

A Portrait of the Steak as a Living Creature

What probably first strikes a contemporary reader in *Gyldendals* 1955 edition is its visually explicit character. Images of dead animals and animal parts constantly remind the reader of the animal source of the meat used in the recipes. Colored photographic charts that display dead animals abound, and are in many cases intermingled with regular recipes. For instance, facing the recipes *Beef and Onion Casserole* and *Beef Stew* are three color plates that display butchered halves of calf, pig and lamb, as well as quarter parts of cattle. The animals hang from hooks to tenderize. The pig still has its head attached, while the cow, the calf and the lamb do not. Black lines drawn onto the photo indicate how the animal is awaiting further division into smaller parts, like head, shoulder, belly and leg for the pig and neck, shin, sirloin and rump for the cow.

Finally, there are two plates that show the offal of these animals: liver, heart, tongue, lungs, paunch and intestines. On the facing page there is a black-and-white photograph of a plate of lamb chops served with peas. The publisher was apparently not worried that color plates depicting raw animal innards would seem unappetizing when juxtaposed with lamb chops on a plate, ready to serve. The connection between the animal, the processing of the animal, and the dish made from it is obvious.

Another example of a pictorial display of the animal can be drawn from the chapter on poultry: the images that accompany this chapter convey the practical instructions needed to deal with the dead bird rather than appetizing images of

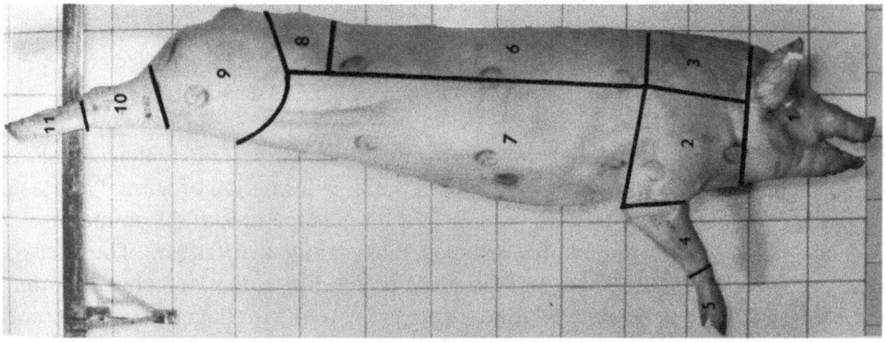

Figure 11.1 In the 1955 edition of *Gyldendals*, an explicit photograph of a pig carcass is delineated to show the various cuts of meat

poultry recipes. A series of five photos demonstrate to the reader how to pluck a hen and burn the remaining hairs from the animal's skin. On the next page, no less than nine photos explain to the reader how to remove the bird's collarbone, chop off its feet, cut off its wings and legs, and pull out its intestines.

The 1955 edition is no less explicit in its textual presentation. The chapter on meat opens like this, in an introduction written by a butcher:

> The term *meat* includes actual meat, blood, and edible offal. These parts of the slaughtered animal have very different nutritional qualities. A normal piece of meat consists of muscles that contain muscle fibers and a greater or lesser degree of fat, ligament and sinews, and possibly also cartilage and bone. (Askevold 1955: 270)

A number of points are made in this short definition. First, it is established that the meat to be used in the subsequent recipes comes from an animal that has been killed to become food—"the slaughtered animal." This is highlighted when the text points to the fact that an animal's body consists of "the actual meat" as well as blood and offal. Lastly, the reader is reminded of the fact that the concrete constituents of meat are the type of thing that one finds even on human bodies; meat is in essence muscles, which consist of fibers, fat, tissues, sinews, cartilage and bones.

The book emphasizes not only detailed descriptions of meat, but of the entire process that makes meat from animals. The nutritional values of the main categories of meat and other animal products are described in detail, and the book explains what the various stamps of the "government's meat inspection agency" stand for. A number of circumstances that are prior to the presentation of pieces of meat in the shop are pointed out as being significant for the cooking process:

> The quality of meat depends on the animal's breed, age, and fodder, how it has been treated during and after slaughter, as well as a number of other factors – some of which are beyond the control of the consumer. (Askevold 1955: 270)

In order for the home cook to be successful with the recipes, the text accentuates that the consumer needs to be familiar with the transformation of animal to meat, in addition to having a quite elaborate insight into the features of the animal itself. The age and sex of the animal, for instance, makes quite a difference. The taste of an ox is considered fuller than, and therefore preferable to, that of a cow or heifer, while a boar must not be allowed to reach sexual maturity, or the meat will become rancid due to the hormones involved in the rut.

About poultry, one reads that, "Young cocks and hens that are no more than 1½ years old, can be prepared like chicken. Cocks and hens between 1½ and 2½ years old are only suitable for boiling. When they grow beyond 2½ and 3½ years old, their meat is tough, and only suitable for making broth" (Askevold 1955: 330).

In the section describing the preparation of hare, one reads:

> Young animals are preferable; the older ones have tough and sinuous meat. A characteristic of young hares is that their ears can easily be torn off lengthways, and that it is easy to lift up the skin between the ears. Older animals' skin is attached tighter to the skull. (Askevold 1955: 272)

In this instance, the home cook herself must learn how to judge the age and quality of the animal, in order to prepare it accordingly. If she concludes, after tearing the hare's ears and stretching the skin on its forehead, that this is an animal suitable to be made into a roast, she proceeds with the recipe for roast hare. The whole process, from a fur-clad animal to meat, is included in the recipe. The recipe is illustrated by two images; first a photo of how to skin a hare, by hanging it up on a hook, cutting its skin loose from the neck and paws, and then stripping it, followed by a photo that shows the undressed hare's back being covered with streaky bacon.

The presence of the animal is in no way concealed; almost all of the animal's body parts are mentioned:

> Cut open the hare's paunch and remove its insides. Fill the paunch with juniper twigs and let it hang in cold storage until preparation. The hare usually comes without guts when it is bought, but it is not always skinned. In the latter case, it must be skinned, see the image. [As caption: "When skinning the hare, cut the skin loose at the neck and the paws as shown here."] Chop off the feet and rinse the hare well in cold water. Be careful to remove all silverskin and sinew, particularly the strong sinew running alongside the back. Chop off thighs and legs … To make sure the back does not bend during cooking, a long trussing needle or knitting needle can be inserted through the spinal canal, or the back can be broken at a few points. (Askevold 1955: 293–4)

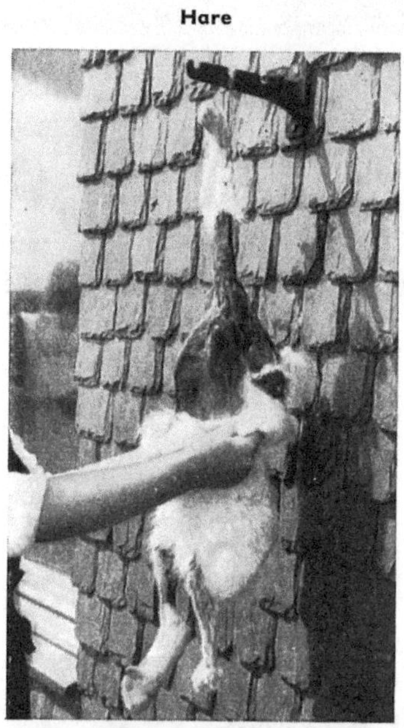

Når haren flås, skjæres skinnet løs ved halsen og ved labbene og trekkes av som her.

Figure 11.2 An unambiguous photograph of a hare being skinned exemplifies the presence of the animal in the 1955 edition

The 1955 edition displays throughout this same consciousness of the whole animal, including the awareness of how the different part of the animal could be used as food. The book contains no less than nine recipes for sausage, and the recipes instructs the home cook in how to use offal, blood, meat and fat to make sausage meat of different kinds—and how to stuff this sausage meat in an animal's intestines. The home cook must know not just what parts of the animal to put into a sausage, but which way to arrange the insides of an animal to achieve the best result:

> If the intestines are not cleaned at a factory, they must be scraped. Place the intestines on a board and scrape the skin clean using a dull knife. If they smell offensive, they should be soaked in water with salt or vinegar. Rinse the intestines well once more. For fresh sausage, the smooth side of the intestines should face outwards; for smoked sausage, the smooth side should face inwards. (Askevold 1955: 324)

In conclusion, the 1955 edition of *Gyldendals* leads the consumer's attention constantly toward the prehistory of meat, to the steak's previous life as a living creature, as it were. In other words, in the 1955 edition, the processing of the food does not start with a ready piece of meat, but with an animal.

Meateriality Goes Mawkish

When comparing the 1955 edition of *Gyldendals* with the most recent, 2002, edition, many of the reminders of the connection between animal and meat have been toned down or have disappeared altogether. In the 2002 edition the color plates of whole animals and offal have been replaced by modestly sketched silhouettes in black and white.

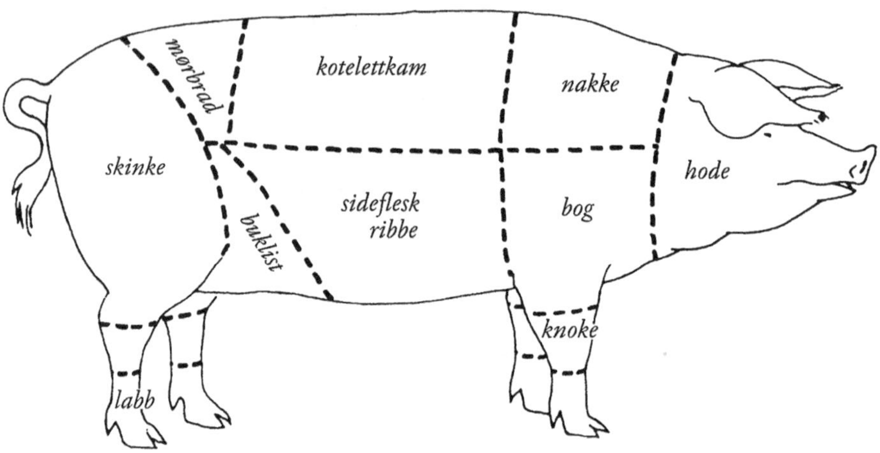

Figure 11.3 In the 2002 edition, the explicit representation of a carcass is replaced by a sketched silhouette of a pig

Beneath these drawings is a schematically organized text that describes the uses of the various pieces of meat. Offal is not included in this schema, and there is no mention of the fact that the quality of meat varies with the age or sex of the animal. With the exception of an ox tongue, the most recent edition contains no color photos of whole animals or of raw animal insides, as the older edition does. The chapter on meat begins:

> Meat is an important source of proteins, vitamins, and minerals. Pure meat is lean food. One can choose among tasty stews and quick dishes made from mince, party dishes made of tenderloin, and marinated roasts of many kinds. (Hovig et al. 2002: 305)

In contrast to the 1955 edition the fact that the meat was previously a living animal is not explicitly mentioned. Nor does the text call forth the same associations to human bodies, since references to muscles, tissue, sinew, cartilage and bone are left out. Unlike in the 1955 edition, blood and intestines are not classified as meat, but as offal, which is seen as a separate category altogether. Furthermore, the introduction to the meat chapter is written not by a butcher but by the book's editor, the famous Norwegian TV chef, Ingrid Espelid Hovig. In summary, meat is described as an appealing ingredient with a useful nutritional content.

The 2002 edition contains many of the same recipes as the one from 1955, but in the former case, the recipes all start with the meat, and not the animal, as raw material. Recipes for hare are included even in the 2002 edition, and a hare is depicted, but the detailed recipe for roast hare, described in the previous section, has disappeared. There are still photos of a hare cut into small parts, and a suggestion to use these bits to make a stew. Compared to many other cookbooks on the market, even this chopped up hare may appear explicit, but in the current context of comparison between the 1955 and 2002 editions, it is obvious that focus has shifted from seeing the animal as the primary raw material, to seeing meat as an ingredient.

A shift toward extended use of ready processed meat is evident. Some of the changes that can be traced between editions may appear minor; for instance certain recipes have disappeared or have been changed. These minor changes might be explained by what it is fashionable to eat at various points of time or technological and material changes, like the appearance of home freezers and ready processed food. Nevertheless, it is interesting to note that there has been a clear shift from representations that explicitly display the connection between animal and meat to representations that only very vaguely uphold this connection.

Sausages are, again, a good example. In the 2002 edition there are six different recipes for sausage stews, where ready-made sausage is among the ingredients (Hovig et al. 2002: 424–7). The content of these sausages is not made clear to the consumer, nor is there any apparent need for such knowledge. Another example of a nebulous meat product is minced meat, which appears in multifarious ways in the 2002 edition. A novel introduction in the 2002 edition is a large color photo of six different types of minced meat: high fat minced pork, low fat minced pork, minced lamb, minced veal, low fat minced beef and higher fat minced beef.

One might wonder why so much space and color is spent on illustrating six almost identical foodstuffs. The minced lamb meat is slightly darker and the minced pork somewhat lighter, but the difference is hardly evident to the eye. Regardless of which animal it origins from, mince consists of muscles that have completely lost their original shape and appearance after going through the mincer. The meat might come from any part of the animal, and whether the animal was old or young, tough or tender, has no impact on the quality.

Figure 11.4 A new entry to the 2002 edition is minced meat in a variety of practically indistinguishable forms

In conclusion, the 2002 edition of *Gyldendals* is still a comprehensive cookbook that aims to inform a modern home cook of all one needs to know to make appetizing and nutritious meals for an entire family. The animal, however, is no longer part of this "comprehensive" whole.

The Fine Line between Carnivorism and Cannibalism

An assumption we made at the start was that every culture needs a strategy for making animals edible. The killing and consuming of animals requires some kind of cultural contextualization that makes these animal deaths acceptable and meaningful. In this section, we put this general assumption into contact with our concrete study of the *Gyldendals* cookbooks.

Our argument is that, in postmodernity, the need for a transformative strategy becomes steadily more acute as the human–animal boundary becomes increasingly porous. To put it differently, the line between carnivorism and cannibalism becomes increasingly thin, and this turns the screw on the need to justify that there is a difference between the two. However, no effective strategy for dealing with this situation is at hand. This is because the blurring of the human–animal boundary is part of a larger process where the old, self-assured notion of humanity and its place in the scheme of things is itself destabilized (Franklin 1999). If "all that is solid melts into air" (Berman 1982), that means that our previously firm notions of human–animal relations also dissolve. This leaves us without a cultural strategy for dealing with the need to kill and consume animals—except to forget, conceal, neglect, overlook. Lacking a generally acceptable cultural framework that validates the killing of animals, we instead make the animal we eat disappear. We make meat a mystery.

The characteristics of our current situation emerge when it is juxtaposed with another cultural context where the human–animal boundary is also felt to be porous, namely hunter-gatherer cultures. Among animistic hunter-gatherer tribes, personhood is not tied to humans. In these cultures, to consume an animal presents a dilemma similar to the one we have described of our own situation, namely the trouble of keeping carnivorism distinct from cannibalism.

For instance, Rane Willerslev's ethnography among the Siberian Yukaghirs describes how, in times of hunger, the Yukaghir shaman can transform himself into an animal of prey, such as an elk, and in that way force the animals to come forth and sacrifice themselves to the hunter. But because the shaman transforms himself into an elk, he becomes a cannibal when he consumes its meat, and as such, he is a non-trusted, dubious figure in the society (2007: 127). Hunters in the same group of hunter-gatherers experience qualms about killing animals for food even when the shaman is not involved, however:

> Although meat is highly esteemed, killing and eating animals is inherently problematic. This is due in large part to the fact that the animals that are most desirable as food – elk, reindeer, and bear – are the ones understood to be most like humans in their moral values and rules of conduct. As one young hunter once expressed it, "When killing an elk or a bear, I sometimes feel that I've killed someone human. But one must banish such thoughts or one would go mad from shame." (Willerslev 2007: 78)

Among some hunter-gatherers, humans and animals inhabit inseparable and interactive worlds. As animals are spiritual beings who relate to both the physical and metaphysical worlds and have interests therein, killing and eating animals are actions with significant moral consequences. Because of this, certain rules and actions—often similar to rules and actions between humans—must be followed in order to maintain the moral order. Because meat stems from beings with personhood, it must be dealt with in an appropriate manner to become edible (Ingold 1994a; Willerslev 2007).

In "From trust to domination: an alternative history of human–animal relations," Tim Ingold sums up the hunter-gatherer worldview by saying that "there is no conceptual distance at all [between humanity and nature], or rather that what we distinguish as humanity and nature, merge, for them, into a single field of relationships" (Ingold 1994b: 18). As we have suggested, when humans and animals are brought onto the same plane in this way, there arises a call for strategies of dealing with the killing and consumption of animals, and Ingold adds, "Given this view of the world, everything depends on maintaining a proper balance in one's relationships with its manifold powers" (1994b: 11). Animals "are not regarded as strange alien beings from another world, but as participants in the same world to which the people also belong" (1994b: 12). Hunting is "a highly successful attempt to draw the animals in the hunters' environment into the familiar ambit of social being, and to establish a working basis for mutuality and coexistence" (1994b: 12). These are the cultural conditions that underlie what Ingold calls *trust* in human–animal relationships, which consists in "an active, prior engagement with the agencies and entities of the environment on which we depend" (1994b: 14). Human–animal relationships are conceived as dialogical, and:

> The animals in the environment of the hunter do not simply go their own way, but are supposed to act with the hunter in mind. They are not just "there" for the hunter to find and take as he will, rather they *present themselves* to him. The encounter, then, is a moment in the unfolding of a continuing – even lifelong – relationship between the hunter and the animal kind (of which every particular individual encountered is a specific instance). The hunter hopes that by being good to animals, they in turn will be good to him by sacrificing themselves to him. But by the same rule, the animals have the power to withhold if any attempt is made to coerce what they are not, of their own volition, prepared to provide. (1994b: 14)

Ingold then goes on to contrast this "single field of [human-animal] relationships" with the highly dualistic view he thinks emerged with domestication. This mode he calls *domination*. The mode of domination consists in:

> the separation of two mutually exclusive domains of being to which we attach the labels "humanity" and "nature." All animals, according to the principle of this separation, belong wholly in the world of nature, such that the differences between species are differences within nature. Humans, however are the sole exception; they are different because the essence of their humanity transcends nature. (1994b: 4)

There is obviously some truth in this. In our culture, the central context in which this mode of domination is expressed, is of course the Judeo-Christian religion. The Bible was in many contexts an unassailable guarantee of human–animal hierarchy. Man alone was created in the image of God; Man named the animals; Man was commanded to subdue the earth (Thomas 1984: 31). Religious tradition thus provided all that what was needed in terms of an animal-to-meat transformation strategy. In the Judeo-Christian tradition, animals were already transformed, as it were, in the sense that they were categorized as belonging to an inferior class.

At a certain remove from traditional doctrine, however, traditional farmer cultures present a context of meaning that could not be described in terms of domination. In fact, a *mutuality* akin to that Ingold describes in the case of hunter-gatherers, was indeed present in the old farmer cultures. Many talk about the traditional farmer's relation to his animals in terms of a contract.

In his classic essay "Why look at animals?", John Berger argued that in industrial capitalist modernity humans have become estranged from all things natural, and are now "a species which has at last been isolated" (Berger 1980: 28). He underscored that this estrangement leads, among other things, to domination. In food production, for instance, animals are "treated as raw material" and "processed like manufactured commodities" (1980: 13). However, Berger also sees that the pre-industrial farmer's relation to his livestock was not one that can be aptly described by that word, and in fact, his article is an elegy on that same farmer's relation of mutuality with his animals. While Berger acknowledges that animals still occupy space in contemporary cultures, he laments the disappearance of integrated, mutually dependent, lives that humans previously lived with animals, where their respective roles were clearly designated and justified. This leads Berger to a seemingly paradoxical conclusion: although animals are superficially present in our cultures, the real animal—the one that entered into serious mutuality with the small farmer—vanishes. In the contemporary age, animals are thus made *invisible*; they are made into objects we can look at but which do not look back: "All animals appear like fish seen through the plate glass of an aquarium" (1980: 16). Although Berger did not comment in particular on contemporary practices pertaining to meat, it is interesting that the concept he uses to describe the contemporary animal's status is that of "vanishing"—which is precisely what we have shown has happened with animals in cookbooks. What has become difficult to accept is the fact that the, "peasant becomes fond of his pig and is glad to salt away its pork" (1980: 7).

The farmer's relation to his animals, as described by Berger, is not wholly unlike that of the hunter-gatherers as described by Ingold. Both are based on mutuality, on a more or less explicit contract envisioned to give each a set of rights and duties. Among hunter-gatherers like the Yukaghirs, this form of sacrificial contract materializes when the elk walks up to the hunter and lets itself be killed, while in traditional Western agriculture, the contract entailed an idea of familiarity: "I will feed you, take care of you, and treat you well, and in the end I will eat you."

The farmer was dependent on his animals, for the food and by-products (like manure) they could provide, and the animals were dependent on the farmer, for grazing areas, fodder and protection against predators and hostile weather conditions. In short, the implicit contract between a farmer and his animals was that they would provide for each other, in turn. Although the contract was implicit, the mutual dependence of humans and animals was explicit, something which is particularly evident in pastoral societies like pre-industrial Norway. Evidently, the mutual dependence of humans and animals was the basis of existence for both. If the animals starved, so did the farmer (Visted and Stigum 1951: 200–3).

While Ingold thinks the aspect of mutuality withers away with domestication, Berger thinks it disappears with industrial capitalism. Regardless of the status of Ingold's more radical claim, one can find additional support for the notion that the industrialization of animal agriculture was a very significant moment in the development of human–animal relations. And the loss of mutuality that Berger laments can reasonably be seen as a result of this process.

Adrian Franklin argues in his book *Animals and Modern Cultures* that the "increasingly contentious and conflictual nature of human-animal relations across a number of sites in the twentieth century" (Franklin 1999: 2) should be seen against a background of the shift toward the postmodern. On the road to describing our present contemporary postmodern situation, he describes what characterized the modern project's conception of animals:

> Whereas modernization was predicated on the essential potential goodness of humanity and built upon the twin goals of progress and democracy, in the latter part of the twentieth century these goals have been deemed unaffordable luxuries of a bygone period ... Animals figured in the modern project principally as a resource for human progress. The destruction of habitat, the enslavement into medical research and the creation of industrial husbandry regimes were perhaps regrettable to those in the know, but they were always justified through their contribution to the great (human) good. (Franklin 1999: 3)

What Franklin refers to as modernization presumably covers some of the same ground as Berger's industrial corporate capitalism. In both cases, the driving force is the technological-social-economic process of industrialization. In this system, animals are conceived of as mere objects, as resources. Even though he refers, again, to the first agricultural revolution (particularly to the domestication of animals), Ingold sums up well the view of animals that pertains under the second agricultural revolution (especially after the industrial intensification of animal husbandry):

> They [animals] become property which can be owned, inherited and exchanged. Property, however, is conceived here as a relation between persons (subjects) in respect of things (objects), or more generally, as a social appropriation of nature. Human beings, as social persons, can own; animals, as natural objects, are only ownable. Thus the concept of appropriation ... sets humanity, the world of persons, on a pedestal above the natural world of things. (Ingold 1994b: 6)

Franklin is concerned to point out that a significant shift has been in the works during the past few decades—incidentally roughly the period between the different editions of *Gyldendals* we have studied—and he adds, "As the modern project collapsed, so also did the sanguine view of humanity," and instead, "a generalized misanthropy has set in" (Franklin 1999: 3). This tendency toward misanthropy is coupled with what Franklin thinks is the clearest tendency in human–animal relations in the latter part of the twentieth century, namely increasing inclusion of animals. "Almost every book published since the 1970s which reflects on modern relationships between humans and animals urges more restraint, more humanity, more paternalism and protection, more respect for animal life and rights" (Franklin 1999: 105). "Inclusion" in this contemporary sense, however, hardly refers to the mutuality that Berger laments. Franklin refers to it rather as "sentimentalization" and "romanticism," and lists pet keeping and animals in entertainment as the most obvious examples. So rather than a return to the farmer's integrated relationship, rather than a return to a contract, we have—as Berger predicted—a situation where animals have been 'co-opted into the *family* and into the *spectacle*' (Berger 1980: 15).

Arguably, this tendency toward anthropomorphism, which reaches its peak expression in children's culture, is only the tip of the iceberg. Underneath the popular expressions of sentimentalism directed at animals—which includes amusement parks, zoos, movies and the like—are more serious attempts to destabilize the long-standing distinction in Western culture between humans and animals. Such challenges come from a variety of sources, not all of them as "sentimental" and "romanticized" as Franklin argues. They come, among other places, from philosophy—from books like Peter Singer's *Animal Liberation* (1975), the so-called bible of the animal movement, or Tom Regan's *The Case for Animal Rights* (1985), or more recently, Jacques Derrida's *The Animal That Therefore I Am* (2008). It comes from fiction, like the Nobel laureate J.M. Coetzee, who with *The Lives of Animals* (1999) (and *Disgrace* (1999)) created characters who spend whole books pondering our relations to other animals, or Jonathan Safran Foer, who switched from fiction to nonfiction to raise consciousness of the animal question, in *Eating Animals* (2009). Finally, such challenges have increasingly come from the scientific community, not least from the books of Donald Griffin, *Animal Thinking* (1984) and *Animal Minds* (1992).

So Franklin is not exaggerating when he says that, "In the second half of the twentieth century … [the] anthropocentric, human progress-orientated view of animals began to change … [and] we start to see the consolidation not only of sentiments, but also of an emotional content and concern not hitherto apparent" (Franklin 1999: 6). One understands more of our current tensions toward animals if one recognizes that anthropocentrism is giving way, slowly and across multiple sites, to biocentrism.

However, this new condition of biocentrism is accompanied by a conspicuous lack of transformative context or ritual. There are no cultural mechanisms or strategies that adequately make killing and consuming animals meaningful. Both the hunter-gatherer cultures *and* Judeo-Christian faith (as well as the everyday practice of farmers living with and in that faith) employ strategies to make sense of animal deaths. In order to make the killing and consumption of animals acceptable and meaningful, they appeal either to motives, to rituals, or to texts emanating from Divinity, and those make the animal edible, by creating a context in which killing and eating a living creature is something other than (or more than) a lust for blood.

Such strategies are what our empirical study—roughly in line with Berger's theory of the vanishing animal—suggests that we have lost. Or perhaps, if we think of neglect and concealment as our contemporary versions of these strategies, it suggests that what is characteristic of our strategy for dealing with killing and consuming animals is that it consists in the lack of one. We have no motives, no rituals, and no book of Divinity with which to validate and justify animal deaths. We have no fish trap. Instead, we respond to the dilemma by removing, neglecting or disguising both the animal that becomes our meat, as well as the site and the process where that transformation from animal to meat takes place.[3]

The Challenge of Rotting Gracefully in the Ground

If the broad cultural tendency we have called biocentrism explains why we feel a need to conceal animal deaths—if it, as we have suggested, creates a (troublesomely) thin line between carnivorism and cannibalism—that is still only one side of the coin. If biocentrism means to grant animals a moral standing similar to that of humans, we have yet to consider the consequences on the other side of that equation. As we have shown, biocentrism entails upgrading the status of animals, in the sense that they are increasingly thought of as morally considerable beings. This tendency is quite ubiquitous in modern Western societies, and arguably leads to many tensions relating to human–animal relations. But the equalization of humans and animals also entails certain changes in our own self-image as a species. Most notably, it stands for a downgrading of humanity—what Franklin refers to as "misanthropy"—which contrasts sharply with the self-image that our civilization grew up with. Human beings are no longer the "crown of creation"; they are—like all other animals—just bodies that wander the Earth for a short period, only to be left in the ground to rot. Robbed of a soul to be redeemed, and with it the prospect of an afterlife in the thousand-year kingdom, human beings are mere matter. The unsettling message that biocentrism presents to a species which has thought of itself as unique is thus: do not think you are special. This is not so much a fall from grace as it is a fall from the *possibility* of grace. And the landing is rough.

In conclusion, perhaps we need to consider whether and how the need to deny (human) death, which many scholars have thought characterizes our time (Becker 1973), is connected to the trouble we have with seeing, admitting and accepting animal deaths. Perhaps our unease with death—in general—has to do with the existential quandary of rotting gracefully in the ground, our stomachs full of beings very much like ourselves.

---

3 Lately some people, not least some prominent leaders of gastronomical opinion (see for instance Fearnley-Whittingstall 2008 and 2011; Oliver 2005 and 2007; Viestad 2010) have attempted to reintroduce the edible animal into our cultural consciousness, and to re-establish the connection between animal and meat—a phenomenon some have dubbed "meat nostalgia" (Parry 2009). Meat eating is justified, say the proponents of this line of thinking, by an awareness of what we are eating and by our producing it knowingly and responsibly. Cookbooks within this particular genre show explicit photos of bleeding animals (see for instance *Jamie's Italy* (2005)), a return to detailed descriptions of how to process the whole beast into delicious meals (see for instance Hugh Fearnley-Whittingstall), and essay-like chapters on food ideology (see Viestad). The question remains whether this is a movement which will become popular or if it is a trend that a select few people at the high end of the income scale will indulge in. A parallel return to facing human death is discussed by Michael C. Kearl and Michael Hviid Jacobsen in Chapter 3 of this volume.

References

Askevold, I. 1955. *Gyldendals store kokebok*. Oslo: Gyldendal.

Barthes, R. 2008. Toward a psychosociology of contemporary food consumption, in *Food and Culture: A Reader*, edited by C. Counihan and P. Van Esterik. New York: Routledge, 28–35.

Becker, E. 1973. *The Denial of Death*. New York: Free Press.

Berger, J. 1980. *About Looking*. New York: Pantheon Books.

Berman, M. 1982. *All That Is Solid Melts Into Air: The Experience of Modernity*. New York: Simon & Schuster.

Brillat-Savarin, J.A. 1848. *Physiologie du Goût, Ou, Méditations de Gastronomie Transcendante*. Paris: Gabriel de Gonet.

Coetzee, J.M. 1999. *Disgrace*. New York: Viking.

Coetzee, J.M. and Gutmann, A. 1999. *The Lives of Animals*. Princeton: Princeton University Press.

Counihan, C. and Van Esterik, P. 2008. *Food and Culture: A Reader*. New York: Routledge.

Derrida, J. and Mallet, M.-L. 2008. *The Animal That Therefore I Am*. New York: Fordham University Press.

Fearnley-Whittingstall, H. 2011. *The River Cottage Cook Book*. London: HarperCollins.

Fearnley-Whittingstall, H. 2008. *The River Cottage Meat Book*. Canada: Hodder & Stoughton.

Foer, J.S. 2009. *Eating Animals*. New York: Little, Brown and Company.

Franklin, A. 1999. *Animals and Modern Cultures: A Sociology of Human-Animal Relations in Modernity*. London: Sage.

Gell, A. 1999. *The Art of Anthropology: Essays and Diagrams*. London: Berg.

Griffin, D.R. 1984. *Animal Thinking*. Cambridge, MA: Harvard University Press.

Griffin, D.R. 1992. *Animal Minds*. Chicago: University of Chicago Press.

Hovig, I.E., Rasmusson, B., Paulsson, L. and Jardhammar, K. 2002. *Gyldendals store kokebok*. Oslo: Gyldendal fakta.

Ingold, T. 1994a. *Companion Encyclopedia of Anthropology: Humanity, Culture and Social Life*. London: Routledge.

Ingold, T. 1994b. From trust to domination: an alternative history of human-animal relations, in *Animals and Human Society: Changing Perspectives*, edited by A. Manning and J. Serpell. London: Routledge, 1–22.

Oliver, J. 2005. *Jamie's Italy*. London: Michael Joseph.

Oliver, J. 2007. *Jamie at Home: Cook Your Way to the Good Life*. New York: Hyperion.

Parry, J. 2009. Oryx and Crake and the new nostalgia for meat. *Society and Animals*, 17, 241–56.

Regan, T. 1985. *The Case for Animal Rights*. Berkeley: University of California Press.

Scholliers, P. (ed.) 2001. *Food, Drink and Identity: Cooking, Eating and Drinking in Europe since the Middle Ages*. London: Berg.
Sherman, S. 2010. *Invention of the Modern Cookbook*. Santa Barbara: Greenwood Press.
Singer, P. 1975. *Animal Liberation: A New Ethics for Our Treatment of Animals*. New York: New York Review.
Thomas, K. 1984. *Man and the Natural World: Changing Attitudes in England 1500–1800*. Harmondsworth: Penguin.
Vialles, N. 1987. *Le Sang et la Chair: Les Abattoirs des Pays de l'Adour*. Paris: Editions de la Maison des Sciences de l'Homme.
Viestad, Andreas. 2010. *Ekte mat*. Oslo: Cappelen Damm.
Visted, K. and Stigum, H. 1951. *Vår Gamle Bondekultur*, vol. 1. Oslo: Cappelen.
Willerslev, R. 2007. *Soul Hunters: Hunting, Animism, and Personhood among the Siberian Yukaghirs*. Berkeley: University of California Press.

Chapter 12
Defacing: Becoming by Killing

Henrik Hvenegaard Mikkelsen

Ritualized beheadings are said to play an important role in male initiation and in molding the configurations of maleness within numerous tribal societies (e.g. Harrison 1993; Cotlow 1953; Ellen 2002: 290). The conjectures about what prompts a man's desire to sever another person's head from his or her body are varied and many, and tales of headhunters continue to stir the minds and spark the imagination of both anthropologists and wider audiences. The native headhunter brings about visceral reactions due to his ability to conjure up graphic images of abject horror in the minds of readers. Simultaneously, such stories adhere to the realm of quasi-real ethno-kitsch and thus maintain a legitimate quality as objects of entertainment and fascination. This became particularly evident as the headhunter made his way into the Western popular literature at the turn of the twentieth century; the image of native warriors beheading enemies, ill-fated missionaries and jungle explorers came to fill the readers with equal amounts of wonder and horror. In this chapter, I wish to retain some amount of wonder regarding this ongoing (though increasingly rare) phenomenon that continues to raise the question: what feature of the head elevates it to such a prestigious level as to be the focal point of ritual violence around the world?

Reminding us to avoid committing the "Cartesian error" of seeing people as passively moving through a stream of objective time, James Weiner (2001) writes: "We make time first through the bodily, spatially situatedness of our goals and postures in life, and it is part of our groundedness to be forward-looking the anticipatedness of these goals" (2001: 35). Humans thus engage both with and in time. In this chapter, I will argue that *ngayó*, a form of ritual beheading found among the Bugkalot of the Philippines,[1] provide the man with a tool to engage with time. The head of the victim was severed from the body and subsequently tossed to the ground with all the force that the cutters, *dima memomotog* (lit. 'the ones who cut') could muster. As I will argue, this treatment of the head (or rather, the face) as a "thing" was a key stepping-stone in the socialization

1 For reasons that go beyond the scope of this chapter, the Bugkalot gradually changed their name during the 1970s. They were then known as the "E'gongot" meaning "from the forest/mountain" by the lowlanders and subsequently became widely known as the "Ilongot" within the anthropological literature through the extensive studies by Renato Rosaldo (e.g. 1980, 2004) and Michelle Rosaldo (e.g. 1980, 1983). In accordance with the wishes of my informants, I will refer to them by term Bugkalot.

of the adolescent into manhood—a transformative process that ultimately aimed to render the man socially autonomous. Through a specific, transgressive act the Bugkalot man stepped out of the unruly realm of the child; no longer dominated by chaotic emotions of shame and anger, he could now accumulate *beya*, knowledge, allowing himself to unfold in time as a person.

If indeed the face is perceived cross-culturally as the essence or surrogate of the person (see e.g. Mckinley 1976; Goffman 1959), its materiality must be of some key importance. Like its Western equivalent, the Bugkalot term for "face," *ganop*, not only refers to a person's physical face. More than describing the various sensory components located on the front of the human head, it also pertains to a certain non-physical feature of one's selfhood that comes into play in social settings. Since, arguably, the face is cross-culturally perceived as the main bodily site of affect, it is highly linked to the social person through rendering visible of what is otherwise a concealed part of the person. The face thus emerges as a key component in a multiplicity of socio-psychological phenomena, such as the human capacity for empathy and the experience of shame. Thereby, a term such as "losing face," that is, a metaphor that applies to the notion of "face" as a social component of one's personhood that one can somehow be deprived of, might not be completely arbitrary. As it was argued by Sartre (1969; also Zahavi 2001), shame means not only being reduced from a fellow human to a mere *thing*. Rather, one becomes an object through the gaze of the Other, who in turn comes to stand out as the subject. Phenomenologically, through the experience of shame one becomes intensely secondary. Shame engendered by experiences of inadequacy is said to characterize the young Bugkalot man. I will establish a connection between the experience of shame of these adolescents and the, until recently, widespread engagement with *ngayó*. I argue that through this ritual murder the young male casts off the social, dependent and thus shameful aspects of his personhood by symbolically removing his own face. Thus, the ritual beheading is tied closely to the transformation from child to adult. But as I will argue in the following *ngayó* should be seen as a form of "suicide," that is, a self-kill, that allows a transformation of the person to take place; not only socially but in the very materiality of his body.

Maleness

The Bugkalot communities are nestled in the Caraballo and Sierra Mountain Range of Northern Luzon. Although the approximately 46 villages are set apart by the expanse of these two mountain ranges, linguistic similarities are found throughout the area. The Bugkalot managed mainly to avoid the process of pacification and colonization that was initiated by the Spaniards and later taken over by the Americans after the Spanish-American war at the end of the nineteenth century. For this reason, unlike most other tribes in the Northern Philippines (see Finin 2005: 107), there were no significant changes in their form of territorial and social organization until the 1970s. Another remarkable aspect of Bugkalot history is that their conversion into Christianity happened very late. It took place as a parallel process with the establishment of socioeconomic relations to the State. This transition was epitomized by the emergence of the Pentecostal and Baptist congregations within the Bugkalot communities and the end of *ngayó* as widespread, openly endorsed phenomenon. Before the church gained a significant foothold up through the 1970s the young men were said to be driven by a desire to "cut," that is, make a ritual kill and thereby become entitled to wear the insignia of manhood: the red hornbill earrings. Nevertheless, *ngayó* has continued to this day though on a much smaller scale (Yang 2011). A friend and primary informant of mine, Tó'gap, whose accounts inform much of the following, committed two murders as late as 1993. And during my last visit in the field in January 2011, in the last stage of my fieldwork, two beheaded men were found in the mountain interior. *Ngayó* has always been an intermittent phenomenon, and though in these years it appears as if the practice is soon to die out, in fact there have been plenty of periods where Bugkalot beheadings were at a minimum (see R. Rosaldo 1980). Nevertheless, changes have taken place and these changes were often articulated to me in the words of my informants, e.g. in the following told me by Tó'gap:[2]

> In the beginning all the men went out to cut [*pamotog*], because that is in our nature. Now that we're civilized and we have heard the word of God, a man is the one who understands the ways of democracy. A man has only good ways [*opeyay ma pengangato*]. A man can always control himself.

This theme of 'control', which I shall elaborate on shortly, saturates most conversations on male ideals.

It is stated by Bugkalots and anthropologists alike, that the Bugkalot society was traditionally organized around an egalitarian power structure with context based leadership. But even today leadership has ad hoc qualities and hierarchies are rarely celebrated. Though "egalitarianism" most often refers to societies of near-equals and, more specifically, a flat power structure (Woodburn 1982), one might argue that it should also be applied to the Bugkalot form of sociality, which is at large guided by egalitarian *ideals*. As within other such forms of society much praise is given to the "autonomy of the individual, and the capacity to act as a free, self-determining, and moral unit or agent" (Kapferer 1988: 15). Thus the egalitarianism is related to the appreciation of a male autonomy, which in this context refers to a certain generic way of talking about the adult, Bugkalot man. This autonomy is engendered in the process of socialization of which *ngayó*, the carrying out of a ritual murder, is said to have been especially effective. My Bugkalot informants talk about this autonomy as the male capacity to act without letting oneself be influenced by the surrounding world. Besides being an inherent attribute of the person that "awakens" in puberty, they claim that the man generates

2 All names are pseudonyms to ensure anonymity of interlocutors.

himself through a process that involves the accumulation of certain defining experiences and acts, which all help constituting him as an autonomous entity.

The anthropologist Michelle Rosaldo (1983), who lived among the Bugkalot in the early 1970s, interpreted these egalitarian ideals as having to do with the way individual autonomy enables the community to minimize internal conflicts. She noticed that the Bugkalot never "require punishment for wrongs, or seek displays of suffering and remorse in making up for untoward violence" (1983: 139). Though few men are as tolerant or self-effacing today, I nevertheless agree with her that in general the adolescent males tend to strive toward not being "different but equivalent or 'the same'" (1983: 137). This, she argues, restrains the development of "conflict-breeding inequalities" (1983: 144). She notices, however, that the Bugkalot do possess a form of ranking, but of the simplest sort: that between women and men, and young and older. The age hierarchy is emphasized in everyday conversation and interaction. She writes:

> Inside the house men *tuydek* (command) women, who in turn pass their commands on to children, and children are quick to *tuydek* those who are younger than themselves. The dynamics of the *tuydek* tend ultimately to permit all adults the orderly poise of the platform – while requests for betel, food, tools, and water set children in almost continuous motion across the relatively unordered floor. (1980: 72)

Michelle Rosaldo explains the Bugkalot impetus toward violence in terms of this structural oppression of the youth:

> In severing and tossing human heads, [Bugkalot] men recount, they could relieve hearts burdened with the "weight" of insult, envy, pain, and grief; and in discarding "heavy" thoughts, they could achieve an "anger" that yields "energy", makes shy and burdened youths "the same" or equal to their peers, and "lightens" both their footsteps and the feelings in their hearts. (1983: 137)

The youth are conceptualized as dominated by emotional, chaotic states and *ngayó*, the kill, is described as having to do with an intense emotion, a cathartic experience, in the sense that it establishes equilibrium and balances one's emotional capacities.

The "Why" of Taking Heads

Michelle Rosaldo notices that certain inherent aspect of male personhood is tied up with the beheading. However she does not answer why. She seems satisfied to observe that local conceptualizations of *ngayó* often revolve around ideas of emotional states, the "heart" (*g'nawa*) and anger or passion (*líget*) (see M. Rosaldo 1980, 1983). Further she argues that *líget* grows within the child and adolescent due to envy (*avet*), that is, *líget* increases when the egalitarian ideals of "sameness" are breached. The act of the beheading is the means for overcoming one's *líget*. Elaborating on Rosaldo's observations, I maintain that *ngayó* have an effect that one does not find in other forms of violent acts. But to understand this one will have to place *ngayó* within the lifeworld of the practitioner, instead of understanding it ultimately from a societal perspective as a customary institution aiming to reproduce certain features of society. The effect of the beheading reaches far beyond any ritual context in which a physical head is required. Most Bugkalot, including Christians, accept that *ngayó* is a tradition transmitted from the time of the earliest ancestors. However, at the same time they have only vague—if any—ideas of how it came into being, or indeed, why they until recently engaged in the taking of heads. The identity of the victim is of no importance. What *is* important is that a human head, or as I will later argue, a *face* is taken. As long as the murder will not cause a conflict within one's own clan, the victim, who is chosen by the elders escorting the raid, is considered appropriate. However, none of my informants were able to explain to me why the head should be severed from the body. The closest I came to an answer was the words told to me by an elderly informant: "To make sure the person is dead. If he is dead he will not come back to take revenge." When I argued that there were other ways to make sure that a person is dead besides cutting of his or her head that part of the conversation came to a halt. Others would argue simply that headhunting was in their nature or that it was an important part of their custom. However, when asked to describe the events surrounding *ngayó* and their individual experiences, another picture emerged. My informants never talked about *ngayó* by reference to spirit beliefs, fertility rites or warfare, but by reference to their individual *desire*; partly a craving after accomplishing the same as their peers, but more often they emphasized that they wanted to cast off the feelings of despair and shame that caused anger and unrest in the young man.

Thus while it has the features of a classic *rite de passage* as it marks the transition into adulthood, what is accentuated is its effect on the emotional life of the killer. This effect goes beyond the *rite de passage*, since it can also be applied in times of anguish or following a loss composure to reduce the *líget* that follows (see R. Rosaldo 2004). In the early 1990s, for instance, my friend and informant, Tó'gap, beheaded a settler to "lift his heart" after having lost his young daughter to illness. However, he stated, though it had a similar effect, this was not the true, customary form of *ngayó*, which should be practiced by the youth. In the following short account, Tó'gap captures the experience of *ngayó* in a way that resonates with dozens of other stories told to me by close informants.

> If you were a young man, you would want to kill. You would want to cut and lift the head and throw it with all your strength. Even though you had walked for days without rest to find the victim, at that moment your heart would be lifted, you would feel brave knowing that this person had been cut under you knife. The *líget* gathered in your mind during all that time … you would no longer feel the

> shame of a child. Your *líget* would vanish and you would no longer fear the gaze of your fellow man. You ask me why and I say because the boy will die. The one that leaves he will never come back.

The first times I heard Bugkalot elders claim that "one did not return from *ngayó*," I thought they referred to the dangers that might be part of the journey. But in fact I never heard of an *ngayó* that had gone wrong. Through Tó'gap's accounts it became clear to me that the statement referred to the fact that something provided this form of ritual murder with the impetus to transform the beheader into another state of being. As Tó'gap's account reflects, this new being represents a motion away from a state associated with shame and "being seen." In general, to understand what have driven young Bugkalots to take heads, one should look into the role of shame among the youth.

Of Men and Mist

Young Bugkalot men often made my fieldwork challenging. When I addressed them they covered their mouths with their hands suppressing embarrassed laughter or stared straight ahead with blank eyes. The reason for this insecurity, this shamefulness, I would come to understand as being closely connected to the self-awareness generated by the gaze of others. This was initially explained to me by Mary, a middle-aged woman in the village:

> [The young men] think they know how to do everything; they *have* to know everything. But they don't and that's the problem, you see? That's why they want to stay out in the forest. They go there to hunt. Or maybe they don't hunt. Who knows? They rarely bring home meat. But they go because in the forest there's no pressure. But back [in the village] when someone talks to them, or even just *looks* at them, it's as if they cannot move!

The language of inadequacy and ineptitude, which permeates much of the everyday talk about the young, is well captured in this quote. The concern with the visibility and invisibility is in fact a recurrent theme in the histories that touch upon the relationship between the young men and the adults. Further, the young men are frequently talked about in terms of instability, inconsistency, translucence and concealment. Terms reminiscent of intangible states—fog, smoke or mist—permeate the various accounts provided by adult as well as adolescent, when they discuss the dispositions of the young men. For instance, as in Mary's account, it is often noted that young men will suddenly disappear from sight as they withdraw to the forest or that they remain quiet, or, at best, make pitiful laconic replies with shameful, downcast eyes when asked even the most simple questions. The man, on the other hand, is said to have overcome his feelings of shame, he can "bear the gaze of others" without fear, since he is unaffected by the surrounding world.

This theme of the young as unbalanced and ethereal shows in the following poem composed by one of Michelle Rosaldo's informants:

> Oh dear, boy, you are as a fog, and all things wait
> dear child, for the moment when you will say the headhunting spells;
> warm your thoughts for the thing you desire, that you
> may, like an airplane, fly to the spirit that you will dismember
> go right on with your plans to kill. (In M. Rosaldo 1980: 141)

Mary explained me during the same conversation that men are supposed to take initiative, to have morals, to be socially active and to be knowledgeable. They should be able to take chances when necessary, to improvise, and travel to far-away places. All these properties she understood to be intrinsically masculine. On the other hand, she said, women like herself were expected to be passive when it came to decision-making, since they acted on emotions and not on *beya*—a form of highly praised knowledge that one generates through one's own individual experiences. The young men do not yet have *beya* and are therefore incapable of making decisions without being influenced by others.

Such requisites seem to place the young men under much pressure and cause a specific negative emotion referred to as *ang-béteng*. This term is derived from the terms *betáng*, meaning a place far away in the wilderness, and *ébtang*, which means "to be separated." Though *ang-béteng* is the feeling of standing outside of the community, of being vulnerable and exposed, it is directly translated into the English word "shame." The young man experiences this form of shame in the confrontation with his own ignorance, fearfulness or incapacity to act independently, or when conditions are publicly exposed. In short, shame emerges when the person experiences a discrepancy between the ideal and the actual. Or rather, when such a discrepancy is being exhibited. The connection between shame and finding oneself within the gaze of the other finds its Western analogy in the writings of Jean-Paul Sartre (1969). He provides a phenomenological analysis of shame in terms of a particular relationship between self and Other. To experience the Other's look is to experience oneself as no longer belonging to oneself but as becoming an object in the project of the Other. The anxiety this may bring about Sartre thus relates to a form of alienation. When someone peeps through a keyhole, he argues, the person is fully absorbed in what he is doing, that is, he is *not* self-aware. However, when he hears a floorboard creaking behind him, he is immediately made aware of himself as an object of the Other's gaze. In this view, shame, the experience of involuntarily becoming an object for the Other, is not a feeling one can elicit on one's own; it requires intersubjectivity, "[it] presupposes the intervention of the other" (Zahavi 2001: 158). Thus, we "appear" in a phenomenological sense, though in an undesirable way, whereby we become estranged to ourselves. In this sense, the young man evaluates himself through the eyes of the Other, and experiences himself not as a person but as a *thing*. This transformation of a person into a thing also appears to be at the core of the ritual beheading, which I will show in the following.

Mocking the Victims

The cutting off of body parts—and heads in particular—is not an ethnographic curiosity of tribal practice. Harrison (2006) has recently described the collecting practices among American soldiers in the Pacific during the Second World War. "Fieldstripping" involved a wide variety of "souvenirs" obtained among the defeated Japanese soldiers and "[the] collection of skulls, teeth, and other body parts seems on the whole to have been carried out as an extension of this 'normal' and widely practiced looting of corpses" (Harrison 2006: 826). Especially, servicemen collected the skulls, though the collectors did not keep them. Instead, the skulls were send home as gifts to friends and relatives. Harrison describes how the practice received international publicity in 1944 when *Life Magazine* issued a full-page "Picture of the Week" with the caption "Arizona war worker writes her Navy Boyfriend a thank-you note for the Jap skull he sent her." The photograph reveals a young woman writing a letter while gazing at a skull placed in front of her on her writing desk. The skull is a love token from a Navy officer in the Pacific (2006: 825). Harrison observes that none of the trophy skulls occurring in the forensic records were collected among German soldiers. They were all identified as Japanese. The corpses of Italian and German soldiers did not undergo the same treatment. Harrison argues that the fetishizing of the Japanese remains as "desirable acquisitions" was encouraged and justified by the racial categorization of the Japanese as having a lower degree of "humanness." This allowed the moral boundaries of normal military fieldstripping to be extended, in the case of the Japanese, to the looting of body parts. However, it was simultaneously through the symbolic act of mutilating the bodies of the enemies that the racialized wartime imagery and conceptions of the Japanese as subhuman were reproduced and sustained (Harrison 2006: 831). Following a common assumption within the Western social sciences, Harrison states that this rendering of the victims into subhumans is a necessary step before such violence can be executed. However, the Bugkalot case offers us an alternative to the notion that recognizing other people in their humanness prevents us from committing atrocities. Seemingly, the Bugkalot found no need to dehumanize their victims beforehand; besides living up to the criteria of *not* being part of the assailant's immediate family, the victims did not belong to a certain category of humans. Nor were the victims classified as "radical others" (Palacios 2004), that is, an "other" with whom there is a relation of antagonism. The victims were Bugkalot and non-Bugkalot alike. But though the victims were not chosen among one's enemies, they were subsequently mocked for their weak and puny character, as feminized objects, in the song *da'eleg'dan*. After the target had been shot and killed from a safe distance and the head had been severed from the body, the victims, who could be men, women and children, would be talked about as "pathetic," *kade-degi*. Ordinarily, this term is applied to women and children and refers to them being dependent on others, and unable to speak their mind because of shame. Thus, the victim is mocked for being part of a human sociality from which the killer has ritually removed himself. This mocking practice might indicate that the prior tossing of the head—and, indeed, the beheading itself—have to do with more than just making a kill; the *memomotog*, the one who cuts, aims to humiliate the victim in an excessive way. This, I believe, has to do with a notion I encountered in several conversions with my elderly informants; Tó'gap explained:

> The *ngayó* is prepared by the Elders. It is their task. They have much experience, so things rarely go wrong. So when someone says that he is not coming back [from *ngayó*], he is not afraid to be killed by the men from the other clan. What's in his mind is that if the *ngayó* is successful he will be celebrated as a *man* when he comes back. Then he will no longer be afraid to speak his mind. Even the Elders will listen to him. The women will find him beautiful since he wears *batling* [red hornbill earrings, the sign that he has taken a head]. When he went out he was a boy. But this boy does not return, you see? What returns is someone else.

The person, according to Tó'gap, was transformed in the act. He not only gains a new status: he in fact is transformed into a different person through a radical, transgressive act.

Transgression

Harrison (1993) asserts that among Papua New Guinean Manambu men, who engage in ritualized murder and cannibalism, a sense of solidarity is formed through "a shared experience of transgressing norms" (1993: 89). Though it would be imprecise to state that my informants were prompted to kill due to a desire for *communitas*—indeed they never expressed solidarity as such to be a driving motive in relation to *ngayó*—there are clear similarities between the Manambu and the Bugkalot. In both cases violence becomes the vehicle through which a sense transgression is reached. Yet, James Weiner (2001) points to an obvious paradox in Harrison's analysis. Weiner argues that "in the moment when a shared way of transgressing norms emerges as a consensual image among men, it becomes as thoroughly normative and conventional as that which it purportedly transgresses." In other words, how can a common, ritualized and widespread practice be considered a moral violation? By raising this question, Weiner also points to a paradox intrinsic to the Bugkalot case. Etymologically "transgression" has been defined as the process of going beyond or infringing a moral principle or an established standard of behavior. But how can a culturally institutionalized and socially encouraged practice be considered subjectively as a moral transgression? Clearly it can be. From an analytical approach that observes the society as a whole, it might appear as an ordinary practice, but for a young man who is standing over his victim clutching a knife in his fist perhaps for the first and only time in his life, this is not experienced as an ordinary event. The task before him is experienced as unfamiliar and even morally problematic. There are contradictions within all societies, which the following account told to me by an elder also shows in a clear-cut way:

> Since I had been attending church at that time, I thought that I would never go on *ngayó*. None of my older brothers had yet taken a head. But then I became a young man and at that time a group of my peers were preparing *ngayó* and I was asked by the elders to join.

Though this brief account does not point toward a transgression as such it seems evident that normativity is not a straightforward matter. Weiner's criticism originates in a perception of primitive society, in which there is little room for internal contradictions. Present-day beheadings are highly surreptitious and few of my informants would openly admit to condone such acts. However, they insisted that such attitudes were not tied to higher degree of moral awareness in the face of modernity and the introduction of Christianity; they had always entertained ambivalent attitudes toward *ngayó*.

Self-evidently beheadings are acts of transgression. But why? And what features of the head elevate it to such a level as to have this transformative capacity? Bataille (1962) has argued that transgressions are never unlimited acts of absolute chaos; they are always constrained within boundaries of time and space; transgressions belong to certain social domains and thereby follow certain rules. And, indeed, the Bugkalot society at large is not violent and the ability of the man to restrain himself is highly valued. It makes it even more interesting, then, that the key *rite de passage* within this tranquil society had to do with young men engaging in a form of ruthless violence, which left even other headhunting tribes in the region astounded.[3] *Ngayó* is transgressive in the sense that it directly goes against the social norms of everyday life; it involves a violation of established boundaries regarding the ordinary treatment of both the bodies of the living as well as the dead. The dead body is a significant entity for a variety of reasons. It is a reminder of our own mortality, which makes us treat it with a great amount of awe. Therefore the treatment of the body during *ngayó* seems surprisingly brutal. The head of the victim would be cut off by the neck, preferably in one determined strike, though some of my informants reluctantly admitted that this is hardly manageable. Afterwards the head was lifted from the ground and tossed as far away as possible. As one might imagine, this had an impact on the person

3 I encountered several men from the Ifugao tribe who commented on the fact that the Bugkalot ambushed their unsuspecting victims and killed without discriminating between men and women, children and grown-ups, healthy and ill. The form of headhunting carried out by the neighboring Igorot tribes had to do with warfare and the ritual treatment of a slain enemy.

carrying out the task. But instead of talking about pity or fear, Tó'gap recalled having become momentarily motionless, transfixed. Not by fear, he claimed, but due to the smell of blood. Only because of encouraging words from the elders leading the *ngayó* was he able to decapitate his victim.

I believe that it is partly due to the victim's humanness—and the radical transgression entailed by the beheading—that *ngayó* obtains its force. However, the particular treatment of the head provides the ritual with a specific impact on the beheader. It is clear that there is something ghastly or grotesque about the severed head, the motionless face, which could be related to its inherent ambiguity; the severed head, which plays the role of a surrogate for the person itself, is somehow de-familiarized, since from being characterized by a constant flow of motion and expressions, it is rendered into an inanimate object—and is treated as a "thing" by being thrown away—while maintaining its human features; it is both a "somebody" and a "something." The severed head thereby belongs to a class of artifacts that seem to radically confound the distinction between persons and things (see Harrison 2006: 831). Since the head, or rather the *face*, plays an important role in so many domains of human interaction, discarding a severed head seems to be a fitting symbol in the Bugkalot man's venture into a state of autonomy.

The face plays a prominent role in a wide variety of scholarly work, from Lacanian psychoanalysis to Erving Goffman's contributions to symbolic interactionism, where the face is seen as a condensed rendering of the person in his or her totality. Goffman (1955) understood "face" as "the positive social value a person effectively claims for himself by the line others assume he has taken during a particular contact. Face is an image of self – delineated in terms of approved social attributes" (1955: 213). However, the most forceful account of the face in relation to intersubjectivity has been put forward by the philosopher Emmanuel Levinas (1969, 1987). Levinas understands the "face-to-face encounter" with the Other as the primordial ethical relation: the encounter with the Other commands us to engage in an ethical relationship. The face is the foundation of ethics since in the face we see humanity in general; thus the obligation to the Other is thereby the obligation to society: "the presence of the face, the infinity of the Other," Levinas famously writes, "is … a presence of the third party (that is, of the whole of humanity which looks at us), and a command that commands commanding" (1969: 213). Though this command concerns an ethical relationship and care for the Other, the face of the Other is not only related to our capacity for empathy; rather the face is the image of sociality at large.

If one accepts the hypothesis that the face is the locus of empathy and the ethical relation, what can this tell us about the role of the head, or rather, the *face*, in certain forms of ritual violence? I propose the following: through the face-to-face encounter with the Other, I see myself in the face of the Other; I realize that, fundamentally, he or she is like myself in my capacity of being human. This has implications in relation to the beheading, since if the Bugkalot adolescent sees himself in the face of the other, whose head, one might ask, does he cut off? By answering "his own" one reaches an insight regarding the impact of the beheading on the beheader. The Bugkalot man thereby not only transgresses the ordinary, ethical state of non-violence of which he is accustomed; he even removes his "face," the childlike, feminine and shameful part of himself, the shared, dependent and social facet of his selfhood. He manifests his selfhood as an autonomous man and will no longer be exposed in social space to the evaluative, objectifying gaze of others. *Ngayó* is therefore considered of defining significance in its bearing on the future life of the cutter. It separates not only the cutter from other children in social space but, perhaps more importantly, violently propels the man out of the chaotic, *emotional* domain of childhood and consolidates the man as a self in time.

Facing Death

The "defacement" marks the commencement of a life trajectory; a trajectory that should, ideally, lead him toward a state of autonomy. This notion of autonomy can be expanded through Heidegger's (1996) thoughts about "being-towards-death." Weiner (2001) argues that this concept refers to the "the self-conscious moment at which humans contemplate their own finitude and undertake the responsibility of finding meaning in these acknowledged limits of human life" (2001: 31). It appears reasonable to propose that *ngayó* is an engine that helps one to reach this form of contemplation. In his magnum opus, *Being and Time* (1996), Heidegger attempts to present an underlying ontology of what he calls human selfhood, *Dasein*. The world is disclosed to us, he argues, as we engage in our "projects." In making a choice regarding a future scheme we thus "project" ourselves forward as the accomplisher of this possibility. Each time we choose, then, as Heidegger writes, "Dasein is already ahead of itself in its being" (1996: 182). Therefore, in some sense we are moving in a circle as we project ourselves forward through our engagement with the world (Jaeger 1958: 58). Unlike other objects in the world, we are not fixed in the present by an inherent being, since this projecting, or "thrownness" (Heidegger 1996: 135), *is* our being. In death, however, this circle ends, which is why the confrontation with our mortality threatens to disclose the very "nothingness" of our being. We face the fact that we can only maintain our being by continously engaging in new future projects. And our death is thereby always ahead of us, always something future that we have not yet realized. "Authentic life," in the thinking of Heidegger, depends on a resolute confrontation with our own being-towards-death: "Once one has grasped the finiteness of one's existence, it snatches one back from the endless multiplicity of possibilities which offer themselves as closest to one – those of comfortableness, shirking and taking things lightly – and brings Dasein into the simplicity of its fate" (1996: 351). Intrinsic to *ngayó* is the moment when the cutter confronts his future demise, the fact that his being is limited in time. Thereby he steps into the world as an, in a Heideggerian sense of the word "authentic" person. The beheader is thereby simultaneously the killer and the dying, and reaches the ideal state of radical autonomy since, as Heidegger claims, "death is essentially, in every case, mine" (Heidegger 1996: 240). Death and subjective autonomy in this view become closely interlinked.

Instruments of Time and the Materiality of the Body

Metcalf (1996) writes that the fact that Bugkalot men did not bring the head back to the village to be used in rituals made him wonder "how contingent the business of decapitation was … might not a knife thrust through the heart have done just as well?" (1992: 274). I have aimed to show that in fact the severing of the head from the body is not an arbitrary act. At the core of *ngayó* one finds a transgressive action linked to the face, to ethics and to one's own death. The shameful child dies in the act and something else takes its place. The autonomy of the cutter is epitomized by his stoic disposition and ability to carry the Other's gaze without showing signs of fear and shame. To this day one will frequently see elders exposing themselves through a certain, highly masculine dance, *ta'gem*, during celebrations. Here, the performer, who is dressed in *aymét*, the traditional male attire, exposes his chest while extending his arms out to the sides, one arm slightly toward the front and the other pointing in the opposite direction—as if the former is holding a shield whereas the latter is preparing to strike a terrifying blow with a sword. The legs are bent, the back is lengthened, and the head is raised, alert. Every muscle is tense, and the facial expression is focused, stern and impervious to the circumstances of the dance, whether it has to do with preparing for a raid or performing in front of spectators. The dancer is accompanied by the *kolesing*, an instrument made from bamboo with rattan strings, played by two persons: the strings is strummed by a man and counterpointed by the sticks played by a woman. As the dancer moves in a vigorous and trancelike manner the staccato sounds provided by the *kolesing* raises and drops in intensity following his steps. The dancer must be untouched by the outside world, decisively confined to himself and his steps. The red color of his headdress is the varnish of braveness, the outward appearance of his ability to be seen. But the cutter has not only changed in relation to the world: he has been physically transformed. Several of my elderly informants recalled that during *magon ma pa'momotog*, the days of cutting, the elders who were "truly knowledgeable" would not die from illness; rather, in the passing of time, their bodies turned increasingly hard, rigid and motionless and became like wood. According to Tó'gap, as the unruly forces of *líget* had been cast off, he would gain control over his emotions and his body. Being untouched by the people around him he did not know of anger and shame. Tó'gap and my other informants associated knowledge with self-possession—the ultimate knowledge takes place when the body becomes motionless. The body of the cutter was transformed in its very materiality. Some were able to remember that these elders at some point would be wrapped in the skin of a boar and placed in a tree in a sitting position. The elder had chosen the tree beforehand. Here the process would continue until, eventually, the body had been eaten by animals and was gone. The ultimate autonomy was thereby found when the bodies turned still. At that point the body would be bathed and exhibited in front of the community. This form of complete, visual exposedness on behalf of the body might be understood as the ideal endpoint of a trajectory of gradual dying that began at the moment of the beheading.[4]

Different forms of bodily materiality have been laid out in this chapter: from the ontologically unsolid body of the youth, to the body which hardens, is condensed in time and space, as *líget* is restrained through the accumulation of knowledge. By treating the Other not as person but as a mere object through a contravention against the empathically conditioned relationship of the face-to-face encounter, the cutter is not only transformed as a social person; he embarks on a process in which he is physically transformed. Interestingly, this conception of the aging body allows for time to be directly tied to, and not something separate from, the man. Temporality is instigated as something originating in his own agency. Unlike the moment of the entering into adulthood, the placing of the frozen, wood-like body in a tree only involved the slightest of ritual activity. This event was merely a continuation of the death that was commenced with the ritual murder and the dramatic reconstitution of the self that it entailed.

Ngayó is an act whereby the subject by engaging in a particular transgressive and intrinsically "anti-social" endeavor, not only demonstrates his existential power to act on the world and thus to be a self-governing, self-generating, "authentic" being. Transgressing against the ordinary ethics and empathic imperatives of everyday life, the very social relation, the act of the beheading is endowed with the force that propels the young man out of his shameful and ontologically unsolid state. This ethical relation I find to be of essential significance in the attempt to understand *ngayó* and, perhaps, headhunting at large. Yet, anthropologists rarely deal analytically with the moral or phenomenological implications of the fact that headhunting actually has to do with one person cutting off the head of not just an (ethnic) *Other*. Scanning through the anthropological literature (e.g. George 1996; Ellen 2002; Hoskins 1996) leaves the impression that besides gaining honor, admiration and social privileges (such as the right to get married) this experience has little impact on the active agent in the particular interaction. My material suggests otherwise; that, indeed, the young Bugkalot man having beheaded another person is transformed in more ways than just as a social person. Due to the empathically conditioned relationship that the beheader violates, the act has an impact on his person in its materiality.

4 Such notions of "gradual dying" as being tied to the very materiality of the body are found in many societies. Bloch (1988) writes that "(there) is nothing mystical about believing … that the process, which leads to what we call death, may begin early in life, perhaps even at death or even before." He notes that among the Merina of Madagascar the child is understood to be wet and soft, but that "the hard and dry in him develops so that by the time he is an old person he is almost totally dry … The completion of this drying takes place only after the burial … Being born, aging, dying, and later burials are all merely episodes with in the same gradual sequence" (Bloch 1988: 13).

References

Bloch, M. 1988. Introduction: death and the concept of the person, in *On the Meaning of Death*, edited by S. Cederroth, C. Corlin and J. Lindström. Acta Universitatis Upsaliensis. Uppsala Studies in Cultural Anthropology, 8, 11–30.

Cotlow, L. 1953. *Amazon Head-Hunters*. New York: Henry Holt.

Ellen, R. 2002. Nuaulu head-taking: negotiating the twin dangers of presentist and essentialist reconstructions. *Social Anthropology*, 10(3), 281–301.

Finin, A.G. 2005. *The Making of the Igorot: Contours of Cordillera Consciousness*. Quezon City: Ateneo De Manila University Press.

George, K. 1996. Lyrics, history, and allegory, or the end of headhunting ritual in Upland Sulawesi, in *Headhunting and the Social Imagination in Southeast Asia*, edited by Janet Hoskins. Stanford: Stanford University Press, 51–89.

Goffman, E. 1955. On face-work: an analysis of ritual elements of social interaction. *Psychiatry: Journal for the Study of Interpersonal Processes*, 18(3), 213–31.

Goffman, E. 1959. *The Presentation of Self in Everyday Life*. New York: Doubleday.

Harrison, S. 1993. *The Mask of War*. Manchester: Manchester University Press.

Harrison, S. 2006. Skull trophies of the Pacific War: transgressive objects of remembrance. *JRAI*, 12, 817–36.

Heidegger, M. 1996. *Being and Time*. Albany: State University of New York Press.

Hoskins, J. 1996. *Headhunting and the Social Imagination in Southeast Asia*. Stanford: Stanford University Press.

Jaeger, H. 1958. Heidegger and the work of art. *The Journal of Aesthetics and Art Criticism*, 17(1), 58–71.

Kapferer, B. 1988. *Legends of People, Myths of State: Violence, Intolerance and Political Culture in Sri Lanka and Australia*. Washington, DC: Smithsonian Institution Press.

Levinas, E. 1969. *Totality and Infinity: An Essay on Exteriority*. Pittsburgh: Duquesne University Press.

Levinas, E. 1987. *Time and the Other*. Pittsburgh: Duquesne University Press.

McKinley, R. 1976. Human and proud of it! A structural treatment of headhunting rites and the social definition of enemies, in *Studies in Borneo Societies: Social Process and Anthropological Explanation*, edited by G.N. Appell. Special Report No. 12. DeKalb: Center for Southeast Asian Studies, Northern Illinois University, 92–126.

Metcalf, P. 1996. Images of headhunting, in *Headhunting and the Social Imagination in Southeast Asia*, edited by Janet Hoskins. Stanford: Stanford University Press, 249–92.

Palacios, M. 2004. On sacredness and transgression: understanding social antagonism. *Culture and Society*, 9, 284–97.

Rosaldo, M. 1980. *Knowledge and Passion: Ilongot Notions of Self and Social Life*. Cambridge: Cambridge University Press.

Rosaldo, M. 1983. The shame of headhunters and autonomy of self. *Ethos*, 11(3), 135–51.

Rosaldo, R. 1980. *Ilongot Headhunting 1883–1974: A Study in Society and History*. Stanford: Stanford University Press.

Rosaldo, R. 2004. Grief and a headhunters rage, in *Violence in War and Peace: An Anthology*, edited by N. Scheper-Hughes and P. Bourgois. Oxford: Blackwell, 150–156.

Sartre, J.P. 1969. *Being and Nothingness*. London: Methuen.

Weiner, J. 2001. *Tree Leaf Talk: A Heideggerian Anthropology*. Oxford: Berg.

Woodburn, J. 1982. Egalitarian societies. *man*, 3, 431–51.

Yang, S. 2011. Headhunting, Christianity, and history among the Bugkalot (Ilongot) of Northern Luzon, Philippines. *Philippine Studies*, 59(2), 155–86.

Zahavi, D. 2001. Beyond empathy: phenomenological approaches to intersubjectivity. *Journal of Consciousness Studies*, 8(5–7), 151–67.

Chapter 13

Dying on Time: Cultures of Death and Time in Muslim Northern Nigeria

Murray Last

In this chapter we briefly turn away from the sociality of death to the actual experiences of the dying and examine how far does the person dying have agency in the timing of their death. Systematic evidence for that agency is hard to come by though there is plenty of anecdotal evidence available. Thus behaviour should offer us here a better guide than belief. So in this chapter some systematic evidence is offered from an old-established Muslim community in northern Nigeria where deaths on a Friday are both preferred and (perhaps surprisingly) achieved: the hypothesis is that to die on a preferred day (or in desirable circumstances) suggests the dying person has some control over the exact timing of his or her death despite cultural taboos that deny it. In short, do not the unusual data offered here imply that some individuals can and do, *of their own accord*, 'turn off' and die?

The cultural context for this Friday preference is laid out but the core question remains: not just how is an excess of deaths on a Friday actually achieved, but how does a dying individual do it? Clearly not all individuals can do it – or more people would, and there'd be no need for elaborate techniques of 'assisted dying'; indeed surely those in extreme pain (even Jesus?) or those being slowly tortured to death would if they could 'turn off' and cheat their tormentors. This 'turning off' is not confined to Muslim Nigerians: for later in the chapter it is pointed out how common it is in non-Muslim cultures anywhere for the ending of a life at a particular moment to appear to be a matter of personal, private choice. But the primary, evidential focus here is indeed on one Muslim culture within which I have lived or worked for over 50 years now. Much has been written over the centuries by Europeans – historians, sociologists, philosophers, theologians – but, naturally enough given their readership, their theoretical preoccupation with life and death, time and its passing, reflect European intellectual assumptions, anxieties and obsessions that spring from a Christian imagination and its morality. The everyday Muslim worlds I lived in simply did not share these intellectual presumptions; dying was taken for granted – in the rural Hausa farmstead where I was a guest for two years 50 per cent of the children I knew were dead by the age of five (65 out of 131).[1] Children's deaths were understood differently from the gradual dying of aged adults or the sudden deaths of robbers and rioters. All in all, death was too commonplace yet too varied

1 Last (1992).

to be elaborated, but the silent grief was often plain to see. One challenge for these four volumes on death, then, is also to look more widely, escaping the theorised 'euro-death' we know and seeking instead to understand the various 'ethno-deaths' ('ethnothanatologies'?) that people have constructed in different contexts. In doing this we cannot assume that every community has, in their everyday practice, as elaborated an 'ethnothanatology' as, say, Christian Europe has (or had).

Indeed, it will be argued here that there is no elaborate 'culture of death' in Muslim northern Nigeria on the scale there is in Christian Euro-America. Death (and resurrection) is a preoccupation of Christians, which is indeed reflected by these volumes: in most churches there is a focus on a Christ dying on a crucifix, and nearby in a graveyard are elaborate family tombs which are tended and swept, even washed, every week. Nor in northern Nigeria is there a 'culture of time' with clocks, watches, bells that marked Europe's early obsession with timing – so strong was the obsession that clocks and death were linked, symbolically implying that the dead were like a clock that had stopped, unwound; even in English today, a heart (with its pulse) can be called a 'ticker', with life 'ticking' away.

My contribution, then, on what happens in Muslim Hausaland springs from my interest as a European, and *not* from my colleagues in northern Nigeria who tolerate but do not approve of my study of dying. Yet I suggest the data I will offer here are significant for the way, despite there being no obvious cultures of either death or time, many ordinary Hausa people behave in dying in a strikingly unique manner. Death may not be dramatised, but some times of dying are seen as quietly better than others – and, most surprisingly of all, some people deliberately seem to achieve death at the 'right' time. This is not 'performed' as Indian or Tibetan gurus are supposed to do but 'naturally': since suicide is taboo in Islam, stopping living just 'happens', they say, as Allah wills it. Personal agency is involved yet concealed – that is the paradox about death and its timing explored here. Its significance ultimately lies not in its being just 'Nigerian' but in its universality.

The Data

It is widely known in Islam that death on a Friday is a blessing given by Allah: it is the best day to die on, if (and only if) He grants it you – not that the Prophet Himself died on that day: He died on a Monday and was buried on a Tuesday, but then the Prophet was exceptional in all He did. The midday Friday prayer at the community's main mosque is the only weekly ritual that should be attended (Qur'an 62: 9–10) by all male Muslims, including boys: so prayers that day are collective, not private. It is at this collective Friday prayer that major announcements are made – for example, in the nineteenth century, instructing the community that a military campaign would start the next day. More importantly, perhaps, it is on a Friday that a trumpet will signal the end of Time; Adam, the first human, was created on a Friday, and therefore presumably died on a Friday too. Friday, however, is not a day off work, nor the first day of the week (which is Sunday), but it is the Muslims' own distinctive day in the way Saturday is for Jews and Sundays for Christians. That dying on a Friday is believed to be 'good' is widely known among Muslims anywhere. But is that just theoretical?

To answer this, in a year-long survey of one traditional cemetery within the old city of Kano in northern Nigeria we found that more graves were dug on a Friday because the gravediggers knew there would be more bodies to bury on that day; and in the course of the year, we found more people did indeed die and were buried on Fridays. Most significant is the fact that as many women and children died on Friday as men (yet it is men who are meant to be the most pious): Allah seemed to bless the very young, even though they were too young to practise their faith, and bless women whose practice of Islam remains hidden within the house. Two questions arose from this finding: first, can one actually choose the moment on which to die (and 'earn' Allah's blessing); and second, if one can, how do the very young (including babies) 'know' it is better to die on Friday?

The following data arose from a study whose start came about by chance. An old friend died in Kano (northern Nigeria) and he was buried, within the ancient city walls, in a major traditional cemetery (*Dan Dolo*) used by the Muslim inhabitants of the local residential quarters – mainly old-fashioned mud-made houses along narrow alleyways. Visiting my friend's grave one day, I chatted to the gravedigger who had identified it for me, and casually I asked him 'how many graves do you dig a day?'. He said he dug most of them himself except on Fridays when others, on their way to work early in the morning, dig a few as an act of piety. 'How many graves in all are dug?' I asked. 'Five to 10 each day, but 15 on Fridays.' 'Why?' 'More people die on Fridays.' My jaw dropped: so the whole of the following year a colleague and I got the gravedigger to keep a daily tally of who was buried when – men, women, children. In Muslim practice, one buries the dead within an hour or two of their death; there is no post-mortem, no bureaucratic paperwork to be done, no undertakers. People in the neighbourhood hear almost immediately, and they will accompany the body on its bier to the cemetery and attend the burial: the graves are already dug. It is a simple, quick ceremony, done in silence. If a person dies in the night, one waits for dawn (there is no electric light). Hence, days of burial are as good an indication as one will get of the actual day of death; there are no official statistics to draw upon later. Even if our statistics have flaws in them (I'd be surprised if they had not), the simple, initial fact remains: five to 10 graves are dug on every day of the week except Friday when 15 are needed.

Dying on Fridays: The Numerical Evidence from Northern Nigeria

There is no statistical evidence, from such states as Britain or the United States that have for years kept good records of dates of death among their populations, to show that any particular day of the week is favoured for dying on.[2] Anecdotally

2 Devis and Rooney (1997).

there may be days that are more advantageous – or simply convenient – for the survivors of a dead patient, but as far as I know there are no Christian or Jewish specially auspicious days for a death.[3] There are diurnal patterns to death, with night/early morning seeing more deaths; in some societies early-morning deaths are so much the norm that people keep a special vigil over the sick, so as to fight off the spirits that may otherwise snatch their patient.[4] But in general, for much of the world, the actual timing of death is not significant: death just happens when it does; it 'comes to you'. Much more important, for most other people, is what or who *caused* the death. But in strongly Muslim northern Nigeria where Allah is the ultimate cause of death, its timing can be unusually significant.

The figures below are based on the daily totals of burials in the cemetery: the gravedigger entered his figures into only three categories – men, women and children; he did not try to assess the age of each child within its shroud. I show Thursday's totals alongside Friday's, as they indicate how almost as significant was a death on a Thursday. Thursday and Friday constitute the traditional 'week-end': Qur'anic and other scholars' schools are closed on Thursdays, so that children and students can have their clothes washed ready for prayers at the mosque on Friday. 'Friday' technically begins at sunset on Thursday evening; it is possible that people just 'missed' dying on a Friday if their death occurred late on Thursday afternoon. Otherwise I don't know why Thursday might be 'special'.

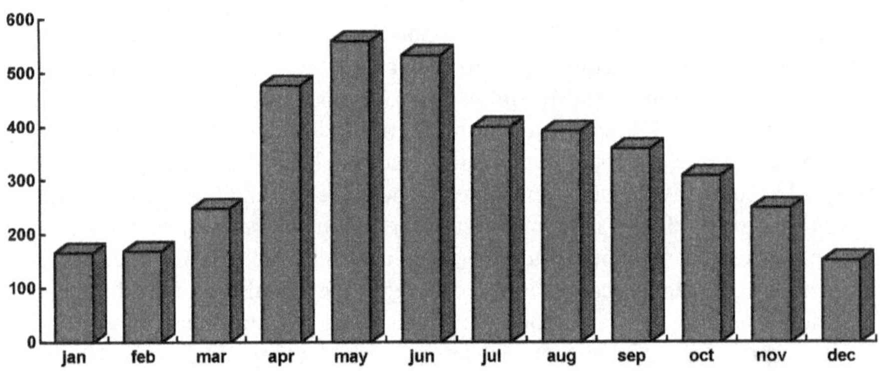

Figure 13.1 Mortality in *Dan Dolo* cemetery

Note: The peaks in April, May and June reflect the measles and cerebrospinal meningitis epidemics usual at this season.

Source: Kano (1996).

3 Apart, that is, from what we are told in Ansu and Ansu (2001).

4 For example, this was done by the Tabwa in eastern ex-Zaire, as reported in Davis (2000).

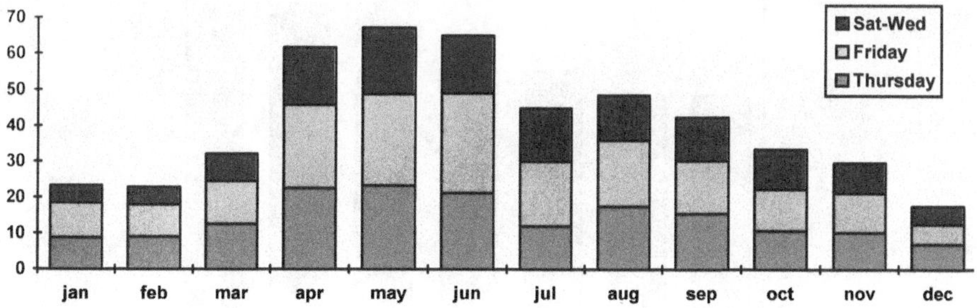

Figure 13.2 Average numbers of burials for each day of the week, averaged over one four-week period (= 'month')

Note: Top segment shows Saturdays to Wednesdays averaged out; middle segment shows Fridays; bottom segment shows Thursdays.

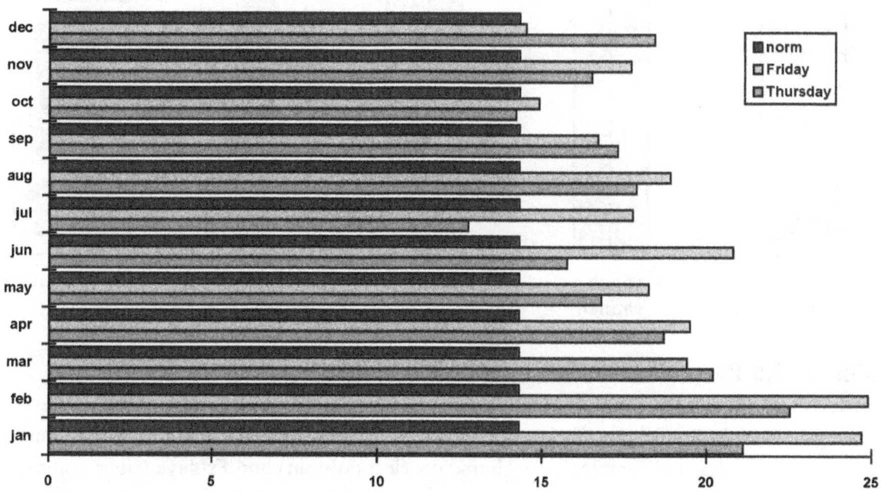

Figure 13.3 Out of all deaths in a month, the percentage of deaths on Thursdays

Note: On Fridays; the norm would be 100 ÷ 7 = 14.3 per cent.

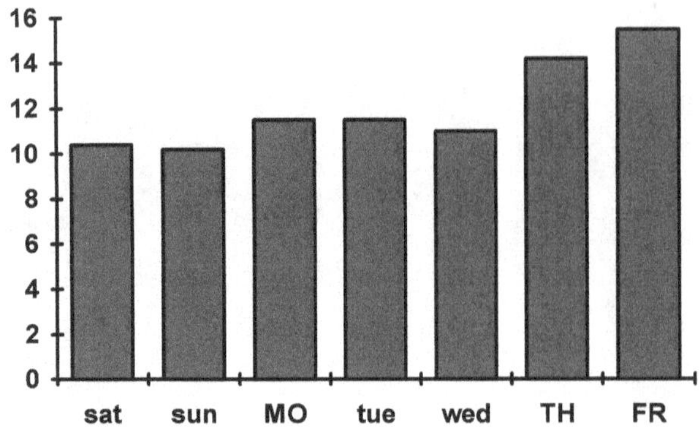

Figure 13.4 Average number of burials totalled by day, for 48 weeks of the year 1996

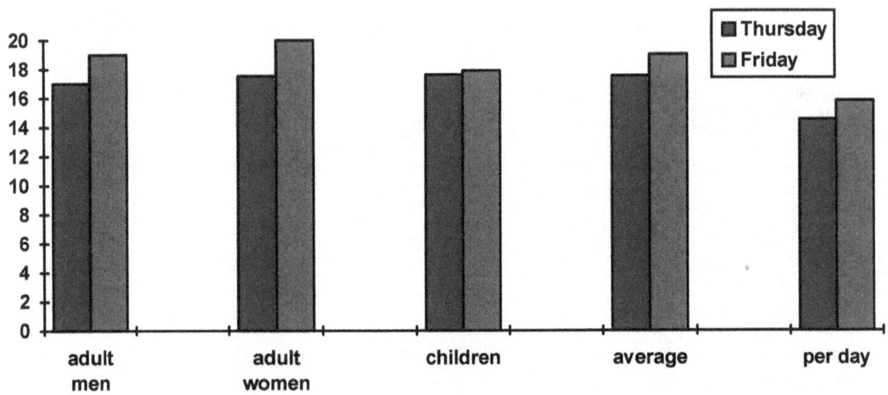

Figure 13.5 Percentage of deaths that occurred over the year (1996)

Note: Amongst all adult males: on Thursdays (left column) and Fridays (right column); amongst all adult females: on Thursdays (left column) and Fridays (right column); amongst all children [males and females]: on Thursdays (left column) and Fridays (right column); average percentages for *all* deaths occurring on Thursdays and Fridays; *real* numbers of deaths occurring on Thursdays (= 14.5 burials), Fridays (= 15.8 burials); the average *real* number for Saturdays–Wednesdays is 10.6 burials.

Responses to these Data

Whenever I have suggested in public seminars in Kano that perhaps the skewing of days of death implies that people may be able to 'choose' in some way when to die, I have met with firm disbelief. It is Allah, and Allah alone, who determines when one dies. Similarly when discussing the data with Islamic scholars (including an imam), I was politely told I was absolutely wrong. It did not help my case when I pointed out (rather playfully?) that it was apparently on women and children, rather than pious old men, that Allah bestowed this particular blessing of a Friday death. It is indeed true that when I searched through such historical records as we have, there was no evidence that any of the famous scholars or emirs of the past were given a Friday death date. In local Kano Muslim culture, my data are, in short, anomalous.

More secular audiences have suggested the higher mortality on Fridays was due, say, to more people coming in to market (or to Friday prayer at one of the mosques) – 'more traffic accidents?', asked some. A more cynical suggestion was that kinsmen store up dead bodies and save them for a Friday burial; another was that hospitals discharge the dying on Fridays, to make for an easier weekend (or something). None of these 'explanations' seemed plausible to me, and there was no evidence for them. What is more interesting to me is that no one apart from the gravediggers knew of this phenomenon. A Friday death is not what everyone is, apparently, trying to achieve; no one seemed worried that their kin had 'missed' the right day to die on. My study was of no interest.

While it does seem possible for some adults at least to 'switch off' much in the same way as one lies down to sleep, if it really was so easy (or well known) more people might presumably do it. It clearly is not easy, and the exact mechanism for how it happens is not clear to medical friends and colleagues. But it is even harder to understand how the very young can do it. One answer is that dying – both for adults but especially for children – in a Hausa household is very largely run by the women of the house. They care for the dying, they will wash the body and wrap it for burial – in which case children will see all this happening and overhear talk about dying on a Friday and its blessings – such as that the Angels of Death will not interrogate the dead on a Friday (they have that day off!) or that all the prayers said by everyone in the Friday mosque will be for the newly dead.[5] In short, a young child grows up knowing this, but what of a baby? My answer would be that the mother–baby link is so close that in traditional medicine one way to treat a baby is to treat the mother, and it is via the placebo route to the mother that the baby responds. Could this be the same with dying: if the mother knows her baby is dying, might she hold it to dying (or somehow 'letting go') on a Friday?

Furthermore, in studies of local-level mortality I did elsewhere, it was indeed the young children who most often died, and young-adult women who were most at risk of death, especially in first childbirth. If death is the province of women, is it then women who are most alert to the blessings of a Friday death, and seek it, consciously or unconsciously, for themselves and for the sick children in their care? In their anguish over a seriously ill child, over a seemingly impending death, does a Friday death offer some sort of recompense?

Whatever the case, there is a calmness in the face of death that is very striking for someone (like me) brought up to make a fuss about dying. It is this calmness that I think may make 'letting go' at the right time – i.e. on a Friday – seem an

5 Cf. Smith and Haddad (1981).

unremarkable yet wise act even for a child in the crisis of an extreme illness: it might even seem 'natural' – an act that can still be carried out even when consciousness is minimal.

'Cultures of Death'

The issue of 'suicide' can be circumvented for Muslims by suggesting that the dying are making themselves ready to die in such a way that death occurs, but its actual occurrence is still at Allah's will. For there is a problem with 'suicide' that is wider than self-killing: it is the notion that someone other than Allah might know the moment of death. The dying therefore do not 'fear' death or how close it is; indeed, from this perspective, my use of the term 'dying' is wholly inappropriate here, since to speak of someone 'dying' would imply one knew their death was very soon. For there is a strong taboo against prognosis: to suggest that this illness or that is leading to someone's death (and therefore the patient *must* be rushed to hospital without delay) is unacceptable. Doctors and hospitals may relieve pain or other symptoms but they do not 'save lives'. Allah *alone* determines the moment (*sa'a*) of death; He has determined one's lifespan (*ajali*) too, but a terminal illness (*ciwon ajali*) is only so called in retrospect, *after* death has occurred: for illness is quite distinct from death – it may accompany death but it is never the cause. Even taking someone seriously ill on a makeshift bed-like stretcher is interpreted as carrying them on a bier to their grave, and is therefore not permitted as it implies you know their death is about to happen; similarly there can be no haste. One may *think* death's very near but one may not act in any way that suggests it. In practice, those near-death, even good Muslims, seem in retrospect to have known it some two or three days ahead and therefore prepared for it, summoning children or sorting out problems in good time. It has been often remarked in Europe how some dogs or cats (even a bird) in a house recognise when a death is near; I presume that it is a result of the changing smell a patient gives off when seriously ill – just as predators can smell a seriously sick animal from afar and come in search of easy prey. The various systems of the body can take time to close down or fail, and the gradual changes are noticeable to the specially sensitive.

Mourning publicly after a death, such as no one going out to work, is restricted by the status of the individual to be mourned: for an ordinary adult or a child a single day of mourning is acceptable, whereas for the head of the household his family take two days off. No noise is made by anyone at the time of death – though I did know one mother who ran out from the house screaming as yet another of her young babies died. Wailing, whether by the graveside or anywhere else, is not done, though the sorrow in people's faces is very visible. When someone died in the middle of the night in the large farmstead I was living in, I was woken by the sudden, total silence: everyone up had stopped what they were doing. Close kin can gather for prayers usually seven days later and again at 40 days but not all do that. Especially important is it for all the dead person's relatives and friends to come in person, preferably on foot, to offer condolences – not necessarily immediately but at some point: writing a letter or a phone call is inadequate, people say. Condoling in person is a simple, quiet act of recognition. In short, any elaboration of death or mourning would be seen as if it was a complaint against Allah.

Anonymity then follows death. Graves in cemeteries are traditionally unmarked – no headstones or the like, just long, low mounds or strips of slightly soft ground reveal a burial; no coffins, no cremation, no urns or ossuaries, no flowers – but you do avoid stepping on a grave or planting over it. There are only a few saints (*waliyai*) who have tombs that can be visited (*ziyara*) for private prayer; politically important men are buried in their family's private gardens or backyards. But 'pilgrimage' is discouraged: the dead are left alone.

Nonetheless, Hausa (like many others who regularly see death first-hand) treat 'death' and 'birth' also as processes, not just as events at a single moment of time. Old age is seen as a process of desiccation of the outer parts of the body where liquids (such a breath or blood) and life flows until they stop; on the inside is the stable, solid self which then ceases to be. After a person has died there are some three days when there is still 'something' there, though the body is buried. So too after a baby is born, it is not clear whether it is truly human – the baby could be a spirit masquerading as human and will 'return' whence it came in a while. Babies, human or 'spirit', are seen to have agency: they are like guests – they have to be coaxed into realising that being alive here is good; if not, they go away, 'dead'. If there are twins in the womb, the 'senior' twin pushes his 'junior' twin out first to test the air: if it is good, the senior twin comes out afterwards – even in the womb foetuses have their own status and power! In short, an individual takes time to be recognised as a human being and at the end of life takes time to be truly 'dead'.

None of this is elaborated (say, into an ancestor cult or an explicit notion of a dead person's soul or spirit) nor is it spoken about, though with very young babies it is more obviously seen in adults' attitudes and wariness of speaking about the newborn. Birthdays, then, are irrelevant, as we do not yet know exactly what has been born. Nor do we know exactly what happens to that stable self once its outer shell (*jiki*) has died: Islamically we are told it is with the corpse in the grave awaiting Judgment Day or being in purgatory (*barzawo*), but privately people, if asked, do not seem so certain. Whatever the case, death itself is only a step in a longer process which only ends for Muslims when Time itself ends. And that may be any day soon.

Anecdotal Evidence from Elsewhere for Choosing the Moment of Death?

'Death is certain but the hour of death is not' [*mors certa, hora incerta*] was a medieval cliché – and still is? – but I wish to suggest in this chapter that the hour, the moment, of death can indeed, in a purely pragmatic sense, be self-determined not through suicide but by the dying person 'switching off', or 'letting go' (as common usage in England has it).[6] Quite what is the biological mechanism that makes this possible is not clear, at least not to the biomedical doctors I have consulted. 'Switching off' is not so easy to do, but it is more common than some of us might think. Anecdotally, however, there is much private evidence that a final 'switching off' of life has often been acknowledged by those at a person's bedside. What is interesting is that even where the cultural rules of a society or a religion in theory forbid such self-termination it happens; and it happens among those we might perhaps least expect it – children, for example.

Some years ago 'near-death experiences' (NDEs) were much discussed, and in each instance the person so near to dying 'came back' or was told to 'go back' as they had 'more work to do' – often children to care for (as last words of the dying, 'I still have more work to do' are not unusual, among the English at least).[7] By being resuscitated, the nearly-dead could of course tell their story. Had they instead 'let go', we'd have no story to hear. These NDEs were primarily made possible by hospitals' development of better resuscitation techniques (that allowed patients almost 25 minutes of apparently 'being dead'), but we have a few other accounts, especially from people who had been in a high fever, which suggests a similar kind of experience – of being told by ancestors or dead kin to 'go back'.[8] Otherwise we have no ordinary evidence of what it is like to be dying, or to become actually dead: accounts of 'resurrection' or 'reincarnation' are not guides to run-of-the-mill experience any more than accounts purportedly coming from the dead via 'spirit mediums'. They may be plausible but scarcely evidence.[9] Hence my data here on 'switching off' or 'letting go' are necessarily indirect and inferred, but are, I suggest, the best we have so far. I hope this collaborative project will offer me more, perhaps better, data.

The problem is not just academic, but could affect policy. For the implication that we are in some sense (and only in some circumstances) in control of our dying – particularly its specific timing – suggests that any drug that inhibits the will or

---

6 The literature on dying and death is huge – two of the most recent studies are Kellehear (2007) and Leader and Corfield (2007); a third is Nuland (1994), but it focuses mainly on hospital practice in America. For dying in Germany, see Eschenbruch (2007: ch. 5).

7 Sabom (1982). Cf. Moody (1975) and Kübler-Ross (1969). The footballer Fabrice Muamba recently spent 78 minutes in London with his heart stopped before he was resuscitated; his experiences, if any, during those minutes have yet to be made public (he is now well).

8 Farnham Rehfisch in Mambila (Nigeria), quoted in Moody (1975); cf. his 1969 essay (307–15).

9 A Ghanaian demographer, Dr Philip Adongo, at a professional conference in Beijing in 1997, has however claimed that his 'study has been the first to be conducted of respondents who are deceased'. I don't think his lead (in using ancestors as direct informants) has been followed up. In Bhutan certain priests claim to have visited the dead on a regular basis, but not for reasons of research apparently. More seriously, an NHS psychiatrist in London, Dr Kenneth McAll, found it useful in some cases to 'heal' a person's ancestors in a ritual as it healed the person too (cf. McAll 1982).

decision to live or to die is more problematic than one might think. For example, what effect does morphine have on the processes involved in dying? As it's not clear what exactly is involved physiologically in 'switching off' or 'letting go', any answer must come from others who are experts, not from me. It is, however, important to know, even if one were to decide that the relief of pain was more important than inhibiting the processes affecting the patient's will to 'hang on' a little longer. Given how some patients 'linger' (to use hospital terminology) beyond the time prognosed for them by their carers and consequently 'block' beds needed for someone else, there is potentially an issue over the timing of death that could be more economic than personal. But does it matter, some might ask – a few extra hours or days (or not) when everyone has said their goodbyes? In short, if the moment of death is a matter of will rather than of God or simply a point in a biological process, might we not need to re-think some of our current assumptions about what is done for, to or with the terminally ill?

Let me start with some of the more anecdotal, everyday evidence. I am sure many readers will have their own stock of anecdotes; whenever I have talked to people, privately or in public, about dying, whether in Nigeria or Britain, in Europe or North America, I have been given personal anecdotes that confirm what I offer here.

The first set of anecdotes refer to the way a dying patient very commonly sends the people around the bedside briefly out of the room, and is found dead on their return. In some instances I know of (in Nigeria for example) their absence from the room has been very brief indeed, and they come back to find 'his eyes weren't his'; his wife held him, but 'his body was quiet'. Hospital staff with experience of being on duty at night have told me this happens very often: the occasion the relative left the patient's room, for a coffee or a wash, was the moment the patient 'chose' to die. Is it a wish just to 'curl up and die' in peace and quiet, as one person said, or a desire not to cause distress to those around the bed? As no one else was present at the moment of death, we do not know how suddenly or slowly the patient 'let go'. Or is it simply that dying people, once alone, can drift off as in sleep whereas if there were others present they might feel the need not to be discourteous and so stay 'awake' for them?[10] Even at the very moment of death do social rules still count? I think they may: whatever it is, 'consciousness' may persist after the more obvious senses have ceased – we know hearing persists after speech and sight have gone, with touch the last sense to go.[11] Obviously not all dying patients die alone – indeed there is a tradition in Muslim cultures that

10 Both Sir William Osler (1904) and Lord Moran (1945: 154), writing of the First World War, reported that the sick and wounded died peacefully. Nuland (1994: 141–3), reports that today the peaceful death is 'a fallacy'. Is it possible, then, that our fear of dying has actually increased over the last century, or just in the United States?

11 On the problem of consciousness, see Humphrey (2006); but much current research is focusing more on how and where the unconscious functions within the brain. Nurses always warn visitors to a deathbed that patients may still hear what is being said.

ideally you hold a patient in your arms and whisper appropriate Qur'anic verses in his or her ear until they die. But some carers, who as Muslims would have done this, did not as they failed to distinguish, they said, between a coma and death, so smooth, so imperceptible had been the transition between life and death. Death did not draw attention to itself.

A second genre of anecdotes concerns close relatives who had, say, a serious heart attack and his or her children were summoned urgently back home, often from so far away it took them a week or two for all to gather round the bedside. Once all were there, the patient seemed very strong and talked happily – but later that evening died. This 'holding on' is recorded for Eskimo Christians waiting for the priest to come at Easter; for others it has been a birthday, or a festival like Yom Kippur or Christmas;[12] others have 'waited' for Ramadan and then died. In one account I heard, the patient, with all her family at last around her, was told 'now you can go', and she was dead within the hour. In some instances, there is a particular date which stands out – for example, the anniversary of a parent's death – and the person chooses (as I'd suggest) to die on that very same day. In other cases there may be a final date by which a person needs to die if she isn't to disrupt totally her whole family's long-awaited project: despite being wholly demented and 'unaware', one such patient nonetheless did die at the last moment she could without causing the expected mayhem. Dementia raises the question, again, of at what level is the decision to 'let go' made. Oliver Sacks, for example, noted how some patients had still an all-but-hidden residual ego, a will to recover, even after years of apparent non-awareness.[13]

In most of the anecdotes I have heard, the speakers' emphasis has been on how the dying 'held on' until the appropriate time; my focus, instead, is on how, once that time came, they 'let go'. But the ability to 'hold on' needs exploring too, albeit in another place perhaps, not least because not everyone in fact can (or do) 'hold on' until their kin come.

A third genre is again told by night nurses of patients who have just had a diagnosis, say of oral cancer, and are due to be operated on the next day: 'they turned their face to the wall', and were dead by morning (this in South Africa). Social anthropologists who have investigated 'voodoo death' – such as deaths 'due' to witchcraft or to a broken taboo where the patient believes he has been ensorcelled or is otherwise fated to die – offer similar evidence, but in these accounts it can take two or three days before the victim is dead: people, the reports suggest, 'give up'.[14]

12 On patterns of dying in Israel, see Ansu and Ansu (2001). Interestingly, many people both in Britain and New York 'hung on' to see the new millennium in: deaths in the first week of January exceeded those in the last week of December by 65 per cent; in New York deaths were up by 50 per cent (Leader and Corfield 2007: 332).

13 Sacks (1985).

14 Lewis (1997). He cites the earlier literature on what became a notorious matter of debate.

In some instances, a person is first declared by his community as 'socially dead' and then led to a place outside to die, whereupon he does die not long after.

In some societies, a failing ruler is told by his councillors that he has to die.[15] He is first given a day to celebrate in special ways, and then at a set time (usually about sunset) sits in a chair, and puts his head back and dies. Some researchers maintain the dying person has taken poison, but no post-mortems were done to establish how exactly death occurred. Other societies, in the past, resorted to more violent ways of ensuring elderly rulers terminate their lives – by burying them alive or shutting them up in a sealed hut: so 'voodoo death' does not occur everywhere, anytime; it is not easy. 'Fear', 'terror' have been said to kill, but the evidence is rarely available except under laboratory conditions (Walter Cannon's 1942 study being the most famous); since then, psychologists have theorised about the mechanics of the autonomic nervous system.[16] The First World War had cases of men apparently dying from shock in the front line, but these, though medically analysed as problems, are still unresolved.[17]

Indian gurus and other holy men who wish to die at the most auspicious time have the appropriate astrological calculations done for them, but we do not know how often they do in fact die at *that* particular special time.[18] Cannot the moment at which they did die always be declared 'auspicious' post mortem? Where I do have an account of a holy person in India choosing to die, she did so by stopping the peristalsis of her gut, and de facto died from the toxins that built up in her intestines, after some two days. Had she, say, stopped herself breathing or lowered her pulse rate, she would have become unconscious or gone into a coma; her personal control of her system would thus have ceased and the body's own automatic mechanisms would have taken over. In short, where people have developed great control over their physical processes, it does not mean they can by that same control necessarily end their lives. It seems that it's not simply the conscious will, the mind explicitly at work, that can 'switch off' a strong, healthy body.

A final genre of anecdote concerns those who know they will die soon, in a day or two; but there is no evidence that they knew or planned exactly when they would die.[19] In one case in Nigeria, the mother summoned her son, her firstborn, and spoke to him in front of others – a sure sign, he said, that she knew she was to die soon, as she had never before in his life spoken to him thus. And she did

15 Soyinka (1975, 1993). They are speaking of the old Yoruba kingdoms in south-western Nigeria.

16 Cannon (1942); cf. Lex (1974).

17 First World War accounts, in Cannon (1915); and the famous account of dead Turks at Gallipoli, cited in Shephard (2000: 3). Cf. Leese (2002). The poet Edward Thomas apparently died from the pressure wave of a shell that passed very close to him.

18 Parry (1994).

19 Michael Sabom (1982: 145–6) makes the point that, while patients knew full well that they would soon be dead, their doctors could not fathom how they knew. On the timing of death generally, see Leader and Corfield (2007: ch. 4).

indeed die that evening. Others I have known knew – and said – they had only a few days more to live, whereas another friend (with cerebral malaria, it turned out, but completely lucid) had no inkling of how close he was to death: he in fact died quietly within half an hour. Other anecdotes tell of people who were convinced they were to die that night and so instigated a ritual of farewell – and then were furious to find themselves awake and alive the next morning. Indeed, a patient sued a hospital for giving him a prognosis of 'a year to live'; he enjoyed spending all his money during that year, only to find he failed to die 'on time'.

A variation on this theme is the so-called 'anniversary reaction' already mentioned where a person 'chooses' to die on an anniversary important to him (like the death of his father); more commonly such people only fall ill 'on time' – they do not die. But Muslim historical tradition has it that a person dies on the same day of the year as he was born; more often I have only found that death has occurred on the same *day* of the week as some significant kinsman's death.

More common are the accounts of people who die within a short time of the death of their long-term spouse or of their leaving home.[20] Alone for the first time in decades, they simply 'give up'; but their death usually takes a while, weeks or months, not days. More sudden was the action of the elderly man I knew who, though sick, was so unwilling to leave his home that he died as soon as he was forcibly put in the ambulance; the ambulance men recognised, they said, the determined look on his face as if he had decided then and there to stop living immediately – despite it being strictly against the rules, they kindly carried him straight back into his house, dead.

Cultures of Time: Days of the Week, not Dates

A study of tombstones in Europe shows how obsessed we are to inscribe on the stone the date, including year, month and number in the month – on which the person buried died. Never is the day of the week inscribed; and this reluctance to name which weekday an event happened is common also among historians as if the various weekdays had no significance. For example, in the US Civil War there was a distinct tendency not to start a land battle on a Sunday except in the western sector and at sea (when ships met by chance?) – but no historian, amidst all the mass of details available, bothers with this sense of timing; Gen. Stonewall Jackson, however, remarked how pleased he was to be dying on a Sunday – I suspect many Christians have felt the same but it has gone unrecorded.

In northern Nigeria, the day of the week on which someone died is more often marked than a date, if only because dates, in a lunar calendar, can vary as to whether the moon was sighted or not. Without printed calendars or diaries, it is the weekday that is important – everyone knows it and both work and markets are allocated by

20 This has been much studied. For a recent summary, see Leader and Corfield (2007: 80–81).

which day of the week it is; indeed the days of the week are named after the market held on that day in the neighbourhood. As a consequence, dates of birth in Muslim northern Nigeria are not traditionally marked, let alone celebrated, and even year-ages matter less; only in the biographies of some particularly famous figures of the past do such calendrical details get recorded afterwards. Traditionally in Islam one's birthday, which is otherwise unrecorded, is determined by one's death-day: the two are meant symbolically to coincide, and ages are calculated retrospectively – so one can never be wholly sure of its accuracy. In practice, however, no one marks the anniversary of a death; one celebrates, for example, only the Prophet's birthday (*maulud*). Death, or indeed the *end* of anything valued, is no matter for special commemoration.

Similarly in the pre-colonial Sokoto Caliphate, the administrative correspondence, written in classical Arabic, was never dated, and indeed many a scholar did not end his book with a date (though some did). Year dates would be calculated from the *hijra* of the Prophet, which was a universal system in the Islamic world, but it seems as if not every scholar knew which exact year he was currently living in: instead, counting often started from a well-known point in local history, like the start of a reign or the beginning of a *jihad*.

While there may not be an elaborate 'culture' of measuring time mechanically or a quest for either chronometric or calendrical accuracy, time in northern Nigeria is nonetheless patterned in ways that everyone can use: it is not dependent on being rich enough to own a working watch or a clock. Thus Muslim farmers in northern Nigeria regulated their day by the sun – the length of a day over the year only varies there by about 30 minutes, so the sun is a constant clock for everyday tasks – the day is divided into blocks of group (*gandu*) or private labour time, and times for the five daily prayers. In the countryside, in contrast to the towns and cities, there is no need for a formal muezzin's call to prayer. The sun's shadow is effective enough as a guide. Similarly it is not a calendar but a change in cloud formations that subtly marks the 'new' (farming) year; different periods of time signal distinct phenomena (such as the annual food shortages, extreme heat or haze, or an explosion of mosquitoes) – few people know which lunar month it is; the major Muslim festivals are geared to the months, of course, but people hear about when they are due – a dying moon is readily observed by all.

On a more general level, it is also worth noting how death and time are metaphorically linked in colloquial usage. Time can die: for example, 'killing time' in English means to while away a span of time waiting for something else that's due to happen, whereas in Hausa 'killed time' (*an kashe mini lokaci*) is when someone has wasted one's time by not, say, turning up on time. Similarly, the Hausa word *lokaci* (time, in the sense of 'moment') can be used euphemistically for death – 'awaiting death' is 'awaiting [one's] time' (*jiran lokaci*), whereas in English 'time' in ambiguous, meaning either 'moment' or 'span' (compare 'its time is up' and 'it's time to go'); time is on the move – it 'passes' – and in Hausa a dead person can be said euphemistically to have 'gone' (*ya wuce*). Thus implicitly death is recognised also a process – the person passes on (if not away). Death, then, is more a separation than a simple cessation of breath in a body. Saying goodbye has started before separation and continues quietly after the actual departure, out of sight. The outer body has visibly been drying up, the senses (similarly on the body's outside – *jiki*) have been failing to register and the individual is left alone (or teased by children). As with any separation, memory of the departed remains, but overt 'separation distress' (of the sort we now diagnose in domestic dogs) is simply not allowed.

In short, spans of time are products of human decision, but the specific moment for anything is determined by Allah alone. We also need to keep notions of 'cause' (of an event) apart from those of its 'occasion': whereas there is a variety of proximate and ultimate 'causes', occasions are singular, spontaneous (like emotions?) – and for us, unpredictable, irrational. Old age, then, is like the slowly withering leaves on a tree: we all see the changes, but the precise point at which a particular leaf falls from the branch is, like death, beyond our ken.

In Conclusion

Something odd, I suggest, is happening here in Kano (and not just in Kano), and it needs to be taken seriously. For on the basis of this evidence I would tentatively argue that there *is* a 'will to die' that can be exercised even under extreme conditions of ill-health. If it is accepted that a dying person can 'hang on', then I suggest there must be a similar ability to 'let go'. But this 'will' is not operative at all times in all persons. I suggest, too, that culturally there is a widespread distaste for anything that smacks of suicide or murder: euthanasia or other forms of assisted death are usually unacceptable. Hence, I think, comes the reluctance, on all our parts, to allow publicly that a dying person can indeed 'switch off': it runs against one of our strongest taboos. I would like to believe that our statistical data from *Dan Dolo* cemetery in Kano strongly support the anecdotal evidence we all seem to have. The fact that the evidence runs counter to Kano's Muslim norms make it all the more remarkable – and that children can express this 'will' to stop living on a particular day is especially important, as in the ethnography of childhood too seldom are children presented as being full actors, as 'agents' in their own right even *in extremis*.

So what other conclusions can we draw from this? Practically I'd suggest that people, especially the dying, should be told of the possibility they can 'let go', they can 'hang on'. Perhaps they should even be taught – as currently American Buddhists are teaching – how to ease the passage into death through controlled breathing. It also challenges the claim by Sherwin Nuland that patients' great anxiety as death approaches is the norm (and not just American), when he totally rejects as a 'fallacy' the calm that Osler and Moran described, back in 1905 and 1915, as common among their dying patients in Britain. Being able to 'turn off' requires a degree of calm in the face of death.

Finally, the comparatively low level of cultural importance given to death in northern Nigeria allows the covert phenomenon of choosing the moment of death to happen 'on the side', almost unnoticed. A Friday death is 'good' but not required by the culture. Our findings about this cemetery in Kano are therefore not seen as significant by local people but they are significant, I suggest, for us because it is a reminder that, whatever our cultures say, one *can* choose the moment of one's end. That is the paradox: our culture's resolute refusal, whether as Muslim Nigerians or Christian Europeans, to countenance the possibility of choosing the moment of our death in fact makes it possible, quietly, to do it. No one expects, no one objects; 'turning off' becomes our final, private act of will.

References

Ansu, J. and Ansu, O. 2001. Death rests awhile: holy day & sabbath effects on Jewish mortality in Israel. *Social Science & Medicine*, 52(1), 83–97.

Cannon, W.B. 1915. *Bodily Changes in Pain, Hunger, Fear and Rage*. New York: D. Appleton.

Cannon, W.B. 1942. 'Voodoo' death. *American Anthropologist*, 44(2), 169–81 [reprinted in *Psychosomatic Medicine*, 1957, 19, 182–90].

Davis, C. 2000. *Death in Abeyance: Illness and Therapy among the Tabwa of Central Africa*. Edinburgh: Edinburgh University Press.

Devis, T. and Rooney, C. 1997. The time taken to register a death. *Population Trends*, 88 (Summer), 48–55.

Eschenbruch, N. 2007. *Nursing Stories: Life and Death in a German Hospice*. Oxford: Berghahn.

Humphrey, N. 2006. *Seeing Red: A Study in Consciousness*. Cambridge, MA: Belknap Press.

Kellehear, A. 2007. *The Social History of Dying*. Cambridge: Cambridge University Press.

Kübler-Ross, E. 1969. *On Death and Dying*. New York: Macmillan.

Last, M. 1992. The importance of extremes: the social implications of intra-household variation in child mortality. *Social Science and Medicine*, 35(6), 799–810.

Leader, D. and Corfield, D. 2007. *Why do People get Ill?* London: Hamish Hamilton.

Leese, P. 2002. *Shell Shock: Traumatic Neuroses & British Soldiers of the First World War*. Basingstoke: Palgrave Macmillan.

Lewis, G. 1997. Fear of sorcery and the problem of death by suggestion, in *The Anthropology of the Body*, edited by J. Blacking. London: Academic Press, 111–43.

Lex, B. 1974. Voodoo death: new thoughts on an old explanation. *American Anthropologist*, 76(4), 818–23.

McAll, K. 1982. *Healing the Family Tree*. London: Sheldon Press.

Moody, R.A. 1975. *Life after Life*. Covington: Mockingbird Books.

Moran, C.M.W. 1945. *The Anatomy of Courage*. London: Constable.

Nuland, S. 1994. *How We Die*. London: Chatto & Windus.

Osler, W. 1904. *Ingersoll Lecture on the Immortality of Man*. Cambridge, MA: Harvard.

Parry, J. 1994. *Death in Banaras*. Cambridge: Cambridge University Press.

Rehfisch, F. 1969. Death, dreams, and the ancestors in Mambila culture, in *Man in Africa*, edited by M. Douglas and P.M. Kaberry. London: Tavistock Press, 307–15.

Sabom, M. 1982. *Recollections of Death: A Medical Investigation*. New York: Harper & Row.

Sacks, O. 1985. *The Man who Mistook his Wife for a Hat*. London: Duckworth.

Shephard, B. 2000. *A War of* Nerves. London: Jonathan Cape.

Smith, J. and Haddad, Y. 1981. *The Islamic Understanding of Death and Resurrection*. Albany: State University of New York Press.

Soyinka, W. 1975. Death *and the King's Horsemen*. London: Eyre Methuen.

Soyinka, W. 1993. *Death and the King's Horseman: A Conversation between Wole Soyinka and Ulli Beier*. Universität Bayreuth: Iwalewa-haus.

Index

Note: Page numbers in *italic* indicate figures.

FSC
www.fsc.org
MIX
Papier | Fördert
gute Waldnutzung
FSC® C083411

Kolibri 360 GmbH
Ferdinand-Jühlke-Straße 7
99095 Erfurt, Deutschland
produktsicherheit@kolibri360.de